MW01342817

THE BOOK OF FLACO

The Book of FLACO

THE WORLD'S MOST FAMOUS BIRD

David Gessner

BLAIR

Printed in the United States of America
Cover design by Laura Williams
Interior design by April Leidig

Owl illustration throughout by Jim DeVona

Blair is an imprint of Carolina Wren Press.

The mission of Blair/Carolina Wren Press is to seek out, nurture, and promote literary work by new and underrepresented writers.

We gratefully acknowledge the ongoing support of general operations by the Durham Arts Council's United Arts Fund and the North Carolina Arts Council.

Library of Congress Cataloging-in-Publication Data
Names: Gessner, David, 1961– author.
Title: The book of Flaco : the world's most famous bird / by David Gessner.
Description: [Durham] : Blair, [2025]
Identifiers: LCCN 2024030756 (print) | LCCN 2024030757 (ebook) |
ISBN 9781958888476 (hardcover) | ISBN 9781958888483 (ebook)
Subjects: LCSH: Flaco (Owl), 2010–2024 | Urban animals—Effect of human beings on—New York (State)—New York. | Captive wild birds—New York (State)—New York.
Classification: LCC QH541.5.C6 G47 2025 (print) |
LCC QH541.5.C6 (ebook) | DDC 591.75/609747/1—dc23/eng/20241115
LC record available at https://lccn.loc.gov/2024030756
LC ebook record available at https://lccn.loc.gov/2024030757

To Hadley

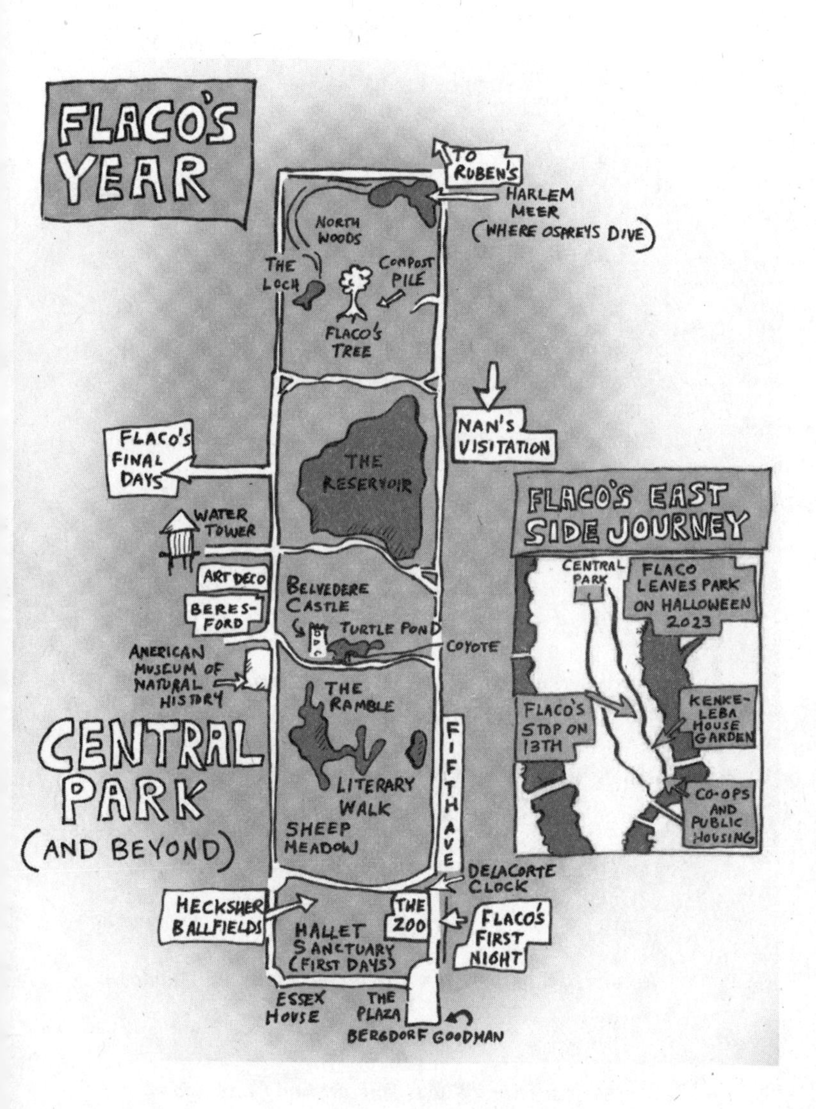
FLACO'S YEAR
TO RUBEN'S
HARLEM MEER (WHERE OSPREYS DIVE)
NORTH WOODS
THE LOCH
COMPOST PILE
FLACO'S TREE
NAN'S VISITATION
FLACO'S FINAL DAYS
THE RESERVOIR
WATER TOWER
ART DECO
BERES-FORD
BELVEDERE CASTLE
TURTLE POND
COYOTE
AMERICAN MUSEUM OF NATURAL HISTORY
THE RAMBLE
FIFTH AVE
CENTRAL PARK (AND BEYOND)
LITERARY WALK
SHEEP MEADOW
DELACORTE CLOCK
HECKSHER BALLFIELDS
THE ZOO
FLACO'S FIRST NIGHT
HALLET SANCTUARY (FIRST DAYS)
ESSEX HOUSE
THE PLAZA
BERGDORF GOODMAN
FLACO'S EAST SIDE JOURNEY
CENTRAL PARK
FLACO LEAVES PARK ON HALLOWEEN 2023
FLACO'S STOP ON 13TH
KENKE-LEBA HOUSE GARDEN
CO-OPS AND PUBLIC HOUSING

Also by David Gessner

A Traveler's Guide to the End of the World: Tales of Fire, Wind, and Water

Quiet Desperation, Savage Delight: Sheltering with Thoreau in the Age of Crisis

Leave It As It Is: A Journey Through Theodore Roosevelt's American Wilderness

Ultimate Glory: Frisbee, Obsession, and My Wild Youth

All the Wild That Remains: Edward Abbey, Wallace Stegner, and the American West

The Tarball Chronicles: A Journey Beyond the Oiled Pelican and Into the Heart of the Gulf Oil Spill

My Green Manifesto: Down the Charles River in Pursuit of a New Environmentalism

Soaring with Fidel: An Osprey Odyssey from Cape Cod to Cuba and Beyond

The Prophet of Dry Hill: Lessons from a Life in Nature

Sick of Nature

Return of the Osprey: A Season of Flight and Wonder

Under the Devil's Thumb

A Wild, Rank Place: One Year on Cape Cod

CONTENTS

An Afterword Before xi

1. The Escape 1
2. Followers 19
3. Evolution 39
4. Visitation 55
5. A Walk in the Park 75
6. The Virtual Bird 101
7. The Passion of Anke 135
8. West Side Story 157
9. Legacy: Flaco Lives! 185

Epilogue: Beyond Flaco 207

AN AFTERWORD BEFORE

I AM STANDING on the edge of a granite cliff, staring up at the owl, who is not twenty feet away, atop the crown of a twisting dwarf pine. He stares directly back at me with eyes that shine orange despite the dying light. I have come 4,500 miles to see this bird, a surrogate for an owl I will never see. Its catlike ears twitch, wind ruffs the feathers on the back of its neck, and as the light dies, the black marks on its buff chest grow more distinct.

You should never expect to win a staring contest with an owl. If you look at it, it will often look back. Not a glance either. A full-on stare. Owls and human beings share forward-facing eyes with binocular vision, and this commonality makes a difference in how we regard them. With them we are face-to-face, eye-to-eye. True, we can't spin our heads fully behind us, 270 degrees, but still we see ourselves in them. What I see now is what I will anthropomorphically call a judging, paternalistic look, the eyebrows slanting sternly over the blazing eyes.

The owl turns away because he has had enough of me. He has other business. His three young are screeching down in the spruce trees below the cliff, a variety of sharp querulous sounds that are all saying the same thing: "Feed me!" He waits though; it is already close to ten at night here near the top of the world, but it is not quite dark enough for hunting. We both notice when the darkness takes another gulp forward, swallowing the light.

Huuhkajat, with its hoot-like first syllable, is the onomatopoeic name for the bird in this land. In my land, we call it the Eurasian eagle-owl, known for its great size and strength. For six months now, I have been reading up on this bird, studying words and pictures about it, and maybe feeling a little jealous when I talked to others about their encounters.

But now it is real. Now I see it. The owl looks beautiful but also bulky and strong. His power is obvious, power in repose. He grips the tree with talons that would not look too out of place on a grizzly bear.

And then suddenly he drops down off the tree, and with two strong flaps of his powerful wings, showing the white of his underwings, crosses by and then below me and down toward the valley below, a dark shape slicing through the gloaming, almost a part of the encroaching darkness, moving toward something that only he sees.

1

THE ESCAPE

THE OWL MAN has a beak.

Of course he does.

Nothing crazy, not too obvious, handsome really. But, to my eye at least, distinctly beak-like. Come to think of it the way he moves his head while keeping his eyes aimed forward is slightly . . . owlish.

I have come to New York City to see the Owl Man and to—this might not come as a surprise—talk owls. Earlier we met at the Belvedere Castle overlooking the Turtle Pond in the middle of Central Park, and we are now making our way through the Shakespeare Garden and over to the west side of the park. The park is the Owl Man's territory, not mine, so he leads the way. Over the next three months this place will take on a mythic quality in my mind, a storybook place like the Hundred Acre Wood, but right now it is still unfamiliar turf.

We emerge from the park around 81st Street and head north, passing the famous Beresford on 211 Central Park West with its massive octagonal towers. A light rain is falling.

The Owl Man's name is David Barrett, and for the last year he has been one of the central players in the drama of Flaco, the Eurasian eagle-owl who escaped from the Central Park Zoo. Almost immediately the escapee garnered a human following, and very quickly a tribe of Flaco followers emerged. If those followers had a high priest, it was David, whose X account, Manhattan Bird Alert @BirdCentralPark, helped others follow the bird both in the park and on their screens.

David is giving me a tour of Flaco's favorite haunts during his last days, like the beautiful art deco building at 241 Central Park West, where, he tells me, "Flaco perched a number of times."

We stop at a spot a little farther north, and he points at another building, brownish beige in color and one row back behind the first line of buildings. There is an exhaust cage on top where Flaco perched.

"This is the last place I saw him," David says.

During his final days, Flaco spent much of his time in hidden courtyards and alleys on the Upper West Side, supplementing his rat diet by occasionally feasting on rooftop pigeons at night. Both, it turned out, helped kill him.

From November 19 to the time of his death this was Flaco's turf, and his hooting echoed down to the streets and through the canyons between the buildings at night.

Now we walk a little farther west, and David points up at a water tower that Flaco favored. I stare up through the scratchy hand of a honey locust that grows out of a dirt patch on the sidewalk of 86th Street. The water tower sits atop a twenty-story building like an ill-fitting chapeau. Its white paint is chipped off in places, revealing a yellow-orange color beneath. An iron ladder climbs up its side, and David describes how Flaco liked to rest on top of that ladder. In fact he had a great fondness for water towers, perhaps because they, like him, perched above everything else on rooftops and because they reflected his hoots, making them echo louder. From up above he hooted down on the city.

I have never really noticed water towers in New York before. After today I will see them everywhere. The great ancient-looking tubs, archaic-seeming, carbuncles from another century topping off modern buildings. Perfect owl perches.

David, who lives on the Upper East Side close to the park, first heard about the owl when he landed on Fifth Avenue the night of his escape, February 2, 2023, David's Twitter followers reporting it to him, "as early as 8:30 or 9." One follower initially reported the owl was a great horned, but David fairly quickly put together the facts that this was a non-native bird and that the Central Park Zoo was only three blocks away. By early the next morning David, armed with his camera and binoculars, had made his way to the Hallett Sanctuary along the park's southern border and soon had the owl in his sights. By then the news was out, confirming the owl's name was Flaco, and he had indeed escaped from the zoo. Over the next few days David watched Flaco, conducted interviews with those who had seen him on the first night, and posted up a storm.

A couple days before meeting him at Central Park, I conducted a long phone interview with David during which he shared many details of Flaco's year of freedom. I noted he had a very precise, almost clipped way of talking, and I suspected he was not born in the United States. But when I asked him about his origins, he answered somewhat vaguely.

"I'm a longtime New Yorker," he said. "The accent is from my training in speech and voice. I was an opera singer for a while, an amateur. I did a lot of classical singing and did work in classical speech, and it sort of stayed with me."

David has been generous with his time and his knowledge, and it is clear how much Flaco meant to him. But in what I am coming to see as his careful, precise way, he does not approach emotions directly.

"He wasn't ever heard hooting in his enclosure back in the zoo," David says as we head back to the park. "Probably because he had no reason to hoot. Owls like to hoot from prominent spots where they can be seen and heard. After he had been free for half a year he was heard hooting from high points, atop water towers and buildings, all over New York. Residents grew used to the hoots and were sad when they were gone."

As he speaks, I try to understand his loss. On a practical level his lifestyle has changed. For almost a full year David lived as a crepuscular and nocturnal hunter. He had always been a—dare I say it—night owl, but with the introduction of Flaco into his life this became more pronounced. Owls are mostly inactive during the sunlight hours, so their workday really starts at flyout, the name for the moment when, after a day of roosting and resting, they launch themselves into the night. But Eurasian eagle-owls like to hunt at dawn and dusk as well as true night, not needing the secrecy of full darkness and surprise that some owls rely on. For the last twelve months David has had his own version of flyout, pushing off from his apartment to the park an hour or two before sunset, armed with binoculars and camera.

David Barrett is, in his own way, passionate, but, as befits a man who studied math at Harvard and MIT, he is also a rational creature of plans, of lists, of goals. Thin and fit, he dresses neatly, and today he wears a windbreaker and the fingerless gloves of a bicyclist. His passion is contained in those fully articulated vowels and clipped consonants, but it isn't hard to see that there is love in his voice when he talks about the owl. There is no other word for it.

The most poignant thing I will hear from David during my visit will not come directly from him but from the playwright Nan Knighton. Knighton was not aware of the Flaco story, when, on November 14, the owl showed up outside the kitchen window of her thirteenth-floor Fifth Avenue apartment. For three hours the owl stayed at her window, an event she would list as one of the most exciting in her life. After this visitation, Nan became a passionate Flaco follower and kept in almost daily contact with David, mostly by text.

When I meet with Nan in her living room overlooking the reservoir in Central Park, she will tell me of her deep fondness for David but also admit that he is a bit of a mystery.

After Flaco had been dead for a week, David made what amounted to, by his standards, an emotional confession to her.

Here is what he said: "I don't know what to do with my nights."

I AM TIRED AND wet when David and I say goodbye, but my day is just beginning. I cut east across the park. My destination now is the Central Park Zoo, and not only for the restrooms. I want to see Flaco's former home, the place he left behind.

Perhaps it is the rain, or David's sadness, or just the general loss, but I find the zoo depressing. Watching my daughter delight in the penguins and sea lions more than a decade ago, it was a different place. But all is context. At this point the dire final necropsy report has not yet cast its shadow over Flaco's year, but it is still hard to hold on to the feel-good aspects of the story that predominated a few months ago. Today the owl has me thinking about freedom and captivity, and so my glimpse of the white flanks of the snow leopard as it passes the glass window does nothing to lift my spirits. The past summer I felt an electric tingle seeing wild grizzlies in Montana and Canada, but today the encaged grizzly slumping on a rock does nothing for me. I remain glum despite the wild avian variety of the tropics exhibit. If I had seen the green peafowl or the plush-crested jay or the golden-crested myna bird in the wild, let alone glimpsed a snow leopard or grizzly from so close, I would have been wildly excited. Here I shrug.

At first I can't find Flaco's enclosure, and I wonder if it

has been dismantled. Over by the back of the grizzly's cage I see what I assume is a maintenance man carrying a ladder. I will give no more details about the man since he will turn out to be the closest I will come to a Deep Throat in my zoo investigation. Staying true to what Philip Roth called the "unseemly profession" of being a writer, I feign ignorance and say, "I heard there was a famous owl who lived here. Do you know where they kept him?"

He says he does and leads me over to the enclosure, which is tucked in near the exit to the penguin and seabird exhibit. I say I heard that the vandals who freed the bird had used bolt cutters to cut the mesh, but he corrects me.

"No, the mesh was steel. They had to use a grinder, a professional tool."

He shows me the enclosure, the mesh now completely removed. He spreads his arms.

"This bird was big, huge, and this was its place," he says.

I am startled by how small the enclosure is. In the notes I took while studying online pictures before this trip, I had written that the cage was about the size of a racquetball court, but it is nowhere close. Barely twenty feet across. It looks like a diorama at a natural history museum. Except it held a live bird.

Now, unprompted, my source speculates about the night of the crime. He suggests it was an inside job.

"They had to know this place. It was somebody who knew the zoo because there was a camera right there."

He points a few feet behind us.

"You step there and you're on camera. They knew to cut the fence in back and come around out of view of the cameras."

I study the space that was Flaco's home for twelve years. Three dead trees that my not-always-reliable phone app identifies as an Indian almond, bagpod, and common fig, served as his perches. An illustration covers the back of the cage, a misty painting of foggy mountains and steppes with a river running through it, Flaco's preferred habitat, put there, if you were in a certain dark mindset, almost to taunt him.

The steel mesh is gone, the former cage now completely open.

I ask my source about this, and he says, "They left it open all year just in case the owl wanted to come back."

As it turns out, he did not.

THERE ARE MANY REASONS that the yearlong odyssey of Flaco the owl captured the imaginations not just of New Yorkers but of millions of people around the world. The formerly caged bird learning street smarts, surviving the mean streets by eating rats. The outsider coming from elsewhere to make it in the big city. The appeal of freedom, of something formerly caged breaking out. And the idea that Flaco, like so many of us after the extended house arrest of COVID, was getting outside and seeing the world. Also the stage on which the drama played out: the island of green in an urban sea that is Central Park, the bird's new home territory. Even the grim necropsy, revealing Flaco had suffered from a viral infection after eating pigeons and had four different rodenticides in his system, serves as a Rachel Carson–esque warning about what we have done to our urban birds.

Over the twelve months between his escape and his death,

Flaco authored a story whose moral, and even its basic narrative arc, is still uncertain and varied, dependent very much on who you talk to or what you read. The poor owl carried so much symbolic weight that it was a miracle he could fly. *Freedom* was the first word on many lips, but human beings, being human, found much to disagree on when it came to the owl. The internet, being the internet, amplified those disagreements. The narrative began with the bird's escape and the zoo's attempts to recapture him, which most supported at first. Once the early predictions that Flaco would not be able to hunt in the (relative) wilds of Central Park were proved wrong, however, the narrative began to change, and there was a growing deep and heartfelt resistance to the attempts to capture Flaco. Free Flaco followers signed petitions urging the zoo to stop their efforts, and their comments flooded the internet, as did criticism of their position. Ornithologists weighed in, some claiming the non-native bird would be a danger to native ones, adding to the theme of Flaco as immigrant, trying to make it in a foreign city. But the experts also worried for Flaco—his ability to survive in the wild, the poisons he might be ingesting once he started hunting successfully—setting up the central conflict of the narrative that would grow over the next year: safety versus freedom.

It is hard to tunnel down through these various narratives and get back to the beginning. But let's try. Try to break through all the various meanings we have pinned on the bird and really see him, not stories about him. Try to see Flaco before the stories begin.

Look into his orange eyes, the color of a monarch butterfly's wings, huge blazing eyes that stare right back into yours. Cartoon eyes but deadly serious ones. Great black pupils. *Human* eyes you might think for a minute, the way they face forward and stare like no other bird, but rounder and wilder than any human eyes you have ever looked into.

Picture a light wind blowing the tufts above the feline ears. When the bird closes those huge eyes his face pinches in toward his sharp, hooked beak. A tight face with little protrusion, disclike. The tawny orange-gold feathers of his full chest are flecked with black, and on his back the striated pattern is reversed: black feathers flecked with orange-gold. He reaches up to scratch himself with great oversized talons, blades that can kill but that he now uses to get a spot under his chin. He hears a twig snap, and his head swivels. He is half asleep one second, fully alert the next. When he hoots, he puts his whole self into it, the white patch below his beak puffing out.

The bird is a Eurasian eagle-owl, given the (silly) Latin name *Bubo Bubo*, *Bubo* being the name for all horned owls in Linnaeus's *Systema Naturae*. A larger cousin of our great horned owl, which is the closest thing we have to an eagle-owl in the Western Hemisphere, they are in fact one of the largest of owls, competitive only with Blakiston's fish owl in size. A predator, born to hunt, which its kind does masterfully. According to Jennifer Ackerman in *What an Owl Knows*, "The most powerful hunter of all owls, it is capable of pretty much taking anything it pleases—rabbits, geese, coots, foxes, even roe deer—surprising prey by flying close to the ground or treetops or seizing birds or bats in full flight."

But this bird, raised in captivity and having had his food provided for him, isn't even ready to hunt for the poisoned city rats he will one day feast on, let alone roe deer. He stares out at the world through a pattern of steel mesh. While the euphemistic human name for this place is an enclosure, it is really a cage. The cage is roughly the size and shape of a department store window display, with a few dead branches to perch on. The view of the sky through the caged roof is obscured by last fall's dead oak leaves, with room to fly from branch to branch but more like hopping than real flight for an animal whose genes know it can soar hundreds of feet in the air, kettling on updrafts like a hawk. The terrain is unvaried, domestic, so mostly the bird inside the cage looks forward through the steel mesh at the one landscape that at least changes, offering variety. This is the landscape, during the day at least, of human faces. He stares at them; they stare back. He does not feed; he is fed. He turns his neck, which he can rotate 270 degrees, and takes in the too-familiar place, the changing faces. A light wind ruffles the boa of white feathers below his neck. He sidesteps down the branch a little with his snowshoe feet and oversized talons.

A human being looking back in at the large bird, now thousands of miles from the taiga and rocky steppes of Europe and Asia and North Africa where its kind evolved, might say it looked bored. But be careful. For most of the twentieth century human beings who studied animals and human beings who used words to describe them were warned not to attribute "human" emotions to animals. Anthropomorphism, the great crime. Only lately has common sense and empathy, bolstered by emerging science, returned to the scene and told us what we knew already. To say an animal is experiencing a certain feeling is not plastering a human emotion on an inhuman thing. It is allowing for the obvious but somehow suppressed fact that we coevolved with creatures like this owl for billions of years before splitting off and going our human way. No grand experiments are needed to conclude that emotion is part of our common heritage. Just watch a big cat prowl back and forth in the zoo. Just ask your dog if it wants to go for a walk.

So maybe *bored* is not the exact word, but something close. Our fellow animal lives a life where many of the things that have been encoded in it by the grand scheme of evolution have been denied. Not small things either. Sex. Food obtained by hunting. Flight. Soaring. Companionship.

Zoogoers sometimes complain about the bird. One day a man tells a zoo worker that the bird looks "grumpy." Perhaps a more worthwhile experiment than questioning the use of anthropomorphism would be to separate this man from his children and place him in an enclosure for twelve years and see how he fares, whether or not he too would exhibit some grumpiness.

This is how life is for our bird, and how life will be forever. To say birds share our emotions is not to say they share all of the evolutionary quirks we have picked up since our families split around six hundred million years ago, particularly the brainy quirks that have developed over the last seven hundred thousand. While it is obvious that we are not the only animals who plan for the future (think squirrels and acorns) and who fear and care for their loved ones (think weeping elephants), we seem to have the market cornered, with the exception of a few chimps, on deep and neurotic pondering of our fates and eventual demise. Perhaps this is a small consolation for our friend in his enclosure. Perhaps. But if he doesn't pass his days in Kierkegaardian contemplation of an endless drab future, he is not unaware of the sameness of his life. If he doesn't know it in his mind, he feels it in his bones. Things will never change.

And then one night they do. Who knows what was in the minds of those who brought about this change, which for the bird will be monumental? Maybe it was two PhDs from nearby Columbia, who had spent the night before drinking coffee and arguing about the ethics of animal imprisonment, synapses sparking while touching on some of the issues discussed above. Maybe just two drunk kids. Or maybe it was a sole emancipator, a heroic loner with a love for birds. An act of vandalism it was called, and vandalism it may have been. Emancipation works, too.

One of the enduring mysteries of that night is that whoever cut the mesh has not been caught. And if they were caught now? At one point they might have been celebrated like ornithological Robin Hoods. Now, in the shadow of the necropsy report, would they be called murderers? (They basically were called this on X, where a petition to catch "the vandal" immediately garnered forty-eight thousand signatures.)

But back to our friend in his enclosure. How long did it take for him to notice that something had changed? How long did he take to approach this changed thing? He would have had no idea at first what this meant, but perhaps soon enough after the interlopers—his liberators it would turn out—left, he flew down to a lower branch to see what there was to see. Hopped farther out on the branch in an awkward graceful way that would make later observers smile.

What he found was an opening, a spot where the mesh was cut and bent back, circular, not too much bigger than an owl. It wouldn't have seemed at first what it would end up being: a portal to another world.

Curious, the owl perhaps turned his head, inspecting in that manner that has led human beings through the centuries to call his kind "wise." And curiosity, another shared trait between his kind and ours, would have finally led him to step or fly or fly-hop through the portal.

To have been in a cage—for twelve years.

To not be in a cage.

Taking care not to anthropomorphize, let's not say he immediately felt "free."

Uncertainty and attendant fear would have likely been the prominent sensations. This was new territory in every sense. Perhaps his first flight after his escape would have been over the waist-high black wrought iron fence that marked the zoo's perimeter. (Any human could step over this after hours, but what usually stops them from doing this, as my friend at the zoo explained, is not the fence itself but the cameras.) This initial flight would have landed Flaco almost face-to-face with his great tormentor, the Delacorte Clock.

Putting fears of anthropomorphizing aside, I don't think I am overstepping by saying that clock must have caused the poor bird torment. During his twelve years in the cage the clock, less than a hundred feet away from his enclosure, went off every half hour during daylight hours, the very hours when any sensible owl would be trying to get some rest. And by saying it "went off" I don't mean it merely rang or gonged. I mean that every thirty minutes the clock went into Broadway production mode, a full-on show that people gathered to see. The clock sits atop an arching passage over the east sidewalk adjoining the park, a three-layer affair, like a brick-and-concrete wedding cake with a rotating stage on the first level on which a cast of animals celebrates every daylight half hour. The hourly notes, even more elaborate, are rung on a large bell atop the clock by two monkeys wielding long-handled mallets, followed by the gong, and then the ringing chorus during which a bear playing tambourine, a penguin on drums, a kangaroo with a French horn, a goat with a flute, a hippo with a violin, and an elephant playing accordion turn in a slow circle. The song and dance of these rambunctious animals would have been a persistent sleep stealer for an animal with ears that can pick up the heartbeat of a vole under a foot of snow.

Once Flaco was over the fence and past the clock, he might have flown toward the rhododendrons near the park's border and perhaps made his way out onto Fifth Avenue through the first human entrance into the park at East 64th. If he had flown in the other direction, west, he would have found the relative peace of some trees and even woods. But he didn't know that; he was flying blind.

As Flaco flew south he would have crossed the street and had his initial encounter with new potential predators: the North American car and its subspecies, the New

York City taxi. His first flights were short, carrying him only three blocks from the zoo, where he landed on the sidewalk around Fifth Avenue and 60th. Landing would prove especially challenging during his first weeks out of the zoo: having never had to land he simply hadn't learned how.

Observers said that even these short flights seemed to exhaust him. Flaco loitered there on the sidewalk for an hour or so, no doubt confused, and it was there he first drew the attention, if not yet the fascination and obsession, of New Yorkers. When Edmund Berry, a birder and photographer, came upon him at 60th Street and Fifth Avenue, he was not sure the owl could fly. Flaco simply sat on the sidewalk while onlookers gawked. Police blocked off the area with yellow tape, and emergency workers tried to capture the bird while police lights flashed in the background. Berry took one picture of the bird looking askance at what appears to be a cat carrier that had been placed on the sidewalk right next to him. For a few minutes it was touch and go; the Flaco story could have ended before it started. But finally the owl flew off.

Among other results, the bird's escape would set off a year of bad puns and dad jokes, and the NYPD was the first to indulge, tweeting: "Well. That was a hoot. We tried to save this little wise guy but he had had enough of his group of admirers and flew off."

It was a cold night, and horns and beeping and general chaos greeted the owl as he flew south past the city bikes and pricey residential apartments and bus and subway stops and a line of horse carriages. Doormen stared, shoppers paused to watch, the bustling crowd briefly stopped bustling. After flying south, he would have encountered his first grass since the park at the Grand Army Plaza with its gaudy gold statue of an angel leading General William Tecumseh Sherman somewhere, perhaps to the burning of Atlanta. Flaco's own march continued through the plaza toward another statue, that of a naked woman holding some kind of fruit basket (the goddess Pomona, it turned out, signifying abundance in a strange tribute to Pulitzer, the newspaper king). There he perched in a tulip tree in front of the Plaza hotel and Bergdorf Goodman, with its faceless mannequins dressed as flappers in a display window about the size of Flaco's recently escaped home.

Edmund Berry had followed the owl south, and he found him up in the tulip tree. Berry would have the honor of being the first of millions of postescape Flaco-watchers. He stared up at the owl for an hour, then another hour. He thought, *Well I'm just gonna stay here. I don't know what else I can do that's so awesome.*

"And it was," he told me later. "It was so incredible. I couldn't leave. It took hold of me."

Flaco spent the cold night in the tulip tree, annoyed, according to Berry, by the twigs that kept getting in his eyes.

At dawn the owl lifted off and flew toward the park. Maybe this was instinct, or maybe he just got lucky, but either way it was a crucial move, away from cars and buildings and toward trees and a murky brown body of water ringed with ice known as the Pond, to a small island of trees that was Hallett Sanctuary. It was a short flight of a few blocks from Bergdorf Goodman to Hallett, the smallest of Central Park's woodlands. Hallett rises against a backdrop of buildings along the south border of the park, still part of the city but separate. Its four acres are thickly wooded, and during those first hours of freedom they would indeed serve as a sanctuary for Flaco. Perhaps this relatively steep forested area spoke to something encoded in him, a familiar landscape he had never seen before, though the steppes and rock ledges of his ancestral home did not have skyscrapers behind them.

Hallett was kept locked that day so that employees of the Wildlife Conservation Society (WCS), which ran both the Bronx and Central Park zoos, could attempt a rescue, but by morning a crowd of bird-watchers, including David Barrett and Edmund Berry, had already gathered outside the gates. They pointed their cameras and binoculars up at the crown of the bare tree where Flaco was getting a taste of the trials of postzoo life, being harassed by a red-tailed hawk and blown almost sideways by the wind on one of the coldest days of the year. Cameras clicked and whirred while the owl stayed silent.

Flaco didn't hunt or take the bait the WCS workers offered. Having been fed the day before, as he had been his entire life, hunger would not yet have gnawed. The next few days, and then weeks, would be marked by a reawakening of instinct, though Flaco's progress was slow at first. His human observers, who would soon multiply exponentially, would note the balky flight, the hesitancy, the difficulty landing, and they would worry for him. Soon enough, however, they were using different words to describe his flight.

Graceful was one.

2

FOLLOWERS

I HAVE COME back to New York to learn about an owl, but at the moment it is a coyote that is on my mind.

What makes the particular coyote I am searching for special, if not unique, is the patch of earth he has claimed as his territory. Not the deserts and mountains of the West that many associate with his kind, or even the suburban lawns of the East where coyotes have grown comfortable. No, this coyote is hiding out in the brambles near the Turtle Pond below Belvedere Castle. Which means that it, and we, are in the middle of Central Park. While coyotes have been seen (and heard howling) in the park before, they are far from a common sight. But just yesterday a bird guide and photographer named Alexandra Wang posted a short video of a coyote, perhaps this one, working its way through the sere grass and phragmites at the west end of the pond near the Delacorte Theater.

It is now two weeks after the death of Flaco. Earlier this morning I met David Barrett at the lookout below the castle. We spent some time chatting while scanning the Turtle Pond, where mergansers and beautiful northern shovelers with chestnut patches on their wings and resplendent green heads floated. I recognized the mergansers myself, but my new friend, a far better birder than I am, provided the name of the shovelers.

Now we—David, two fellow birders, and me—are standing on the concrete path that runs down from the castle, binoculars up and peering over a short fence into the briars and scrub that grow along this side of the pond. One of the birders, an outgoing and passionately emphatic woman named J. P., has been out here all morning, and she wields a thermal monocular that looks like a small telescope. This device can pick up the heat signature of an animal, and she believes it has done just that with our coyote down in the brambles. She confirms this with her binoculars and then points me toward what she sees.

I adjust my binoculars and search, trying to aim them where she wants me to. What I see through the lens could best be described as a blurry Jackson Pollock.

"You see it?" she asks. "You see it?"

"Kind of," I say. I don't want to disappoint her.

When I finally focus, I manage to see an orange patch that I think may be the coyote's back but will turn out to be a stump. That will be the extent of my coyote interaction today, not exactly the wild encounter I'd hoped for.

All our searching and pointing has drawn a little bit of

a crowd, which worries J. P. She has reason to be worried. An owl is one thing, but if a coyote is known to be here, or worse, encounters a person, or, worse still, *bites* a person or a dog, it will surely be curtains for that creature. In the northeastern United States the eradication of predators is mostly a thing of the bloody past, unless of course the predator steps over the line and hurts us or one of our pets.

J. P. wants to distract or disperse the crowd to prevent this from happening, and the way she does this is to stage an elaborate ruse.

"Nothing here," she says loudly. "But there are red-winged blackbirds at the overlook. A sure sign of spring!"

J. P. is understandably excited about seeing the coyote. We all are, David Barrett included. Just the searching for it triggers something surely encoded in *Homo sapiens*, and we thrill to the idea of the possible wild, in this case the possible urban wild. And yet while this thrill is very real, I think that, today at least, this excitement is about more than seeing a *Canis latrans* in Central Park.

I believe there is something compensatory about this feeling. I believe it has grown in part out of a gap, a void.

For the last year my three companions have had a deep purpose in their lives, the very thing most of us lack. Moreover, it was a primal purpose, the kind of purpose our species evolved with, the kind that millions of years of trial and error taught us to get excited about. For the last twelve months their lives have been filled with a passionate pursuit and, better yet, a passionate pursuit of an animal. But just two weeks ago that pursuit ended. While seeing a coyote in Central Park is undeniably a wonder, it can't entirely fill what has been lost. Maybe that's why I sense a dash of desperation mixed in with the thrill.

Something is missing.

FLACO'S FIRST EXPLORATIONS AFTER his escape were short and tentative. He stayed close to the Hallett Sanctuary on the southern edge of the park, perched high amid a small forest of beech, ash, locust, cherry, black oaks, and sassafras, all the trees still February barren except for the occasional evergreen hollies and inkberries. With the branches bare, the first wave of human Flaco-watchers had a clear view of not just the bird but of buildings, like the historic Essex House, that stood behind him. Some of the newer skyscrapers dwarfed the Essex, rising high above the trees where Flaco perched. Red-winged blackbirds, harbingers of spring, were not quite back, but over those first days Flaco encountered his fellow free birds: chittering sparrows and that irksome red-tailed hawk who harassed him. This was clearly a different life than the one in his cage.

But one similarity with his old life was that there were always eyes on him. Many more eyes, eventually, than those that looked in when he was in his cage. Despite the frigid cold of his first free morning, word had gotten out, and a crowd of bird-watchers, many of them Central Park regulars, pointed their binoculars up at the windblown owl. This was just the beginning. For the next twelve months Flaco's fans would follow his every move.

From the start Flaco existed in two realities: the Park and the internet. He was always both an actual bird and a virtual one, followed by foot and on-screen. In this way he mirrored us, our world, and his wildness was a curiously chronicled one, cultivated in decidedly unwild ways. But if those who tracked him wielded cell phones and cameras instead of spears or bows, the primitive thrill of the chase was real.

Even during that first week, narratives had begun to flow and intermix. There was worry about the extreme cold, but that was countered by experts pointing out that some wild eagle-owls called the Arctic their home. By the second day a larger concern was that the bird had made no effort to feed himself, still not tempted by the bait the WCS employees laid out on the ground. The zoo had not been cast as a villain yet. The earliest tweets, including David Barrett's, were concerned with Flaco's safety, focused on the hope that the owl would be recaptured, though soon enough the word *freedom* would begin to creep into the dialogue.

A dramatic three weeks followed, all of it chronicled by David Barrett on Manhattan Bird Alert. He and his followers made note when Flaco hooted among traffic noises on February 6 and when he flew to the south end of the Literary Walk, the elm-lined promenade with statues of Shakespeare and Sir Walter Scott, on February 7. That day he swooped unsuccessfully for a rat. Then, on February 8, he flew back to the zoo and dipped into the crane exhibit for a "drink of water" according to one of the zookeepers. (You may wonder: If an owl could fly in, why couldn't the cranes fly out? The answer is that in zoos some larger birds are flight restricted or "pinioned" from a very young age, their pinion joints removed to prevent flight.)

Those who followed the owl took beautiful pictures and videos in which you could (and still can) stare into the bird's huge and perfectly round orange eyes or watch the feathers on his proud black-flecked chest and ear tufts blowing in the cold February wind as he lifts his wings, stretching while high on a branch, taking mincing steps with enormous taloned feet, and eventually hooting, a sound that rises from his chest as he moves his head forward and the white ruff below his beak puffs out. Those who watched were impressed by his beauty and his sheer size (and would have been more so had *he* been a *she*, females being the larger of this dimorphic species).

His followers worried that he had not yet caught a rat. And they worried he might catch one of the rats laid out by the rescue workers to trap him, though at that point most of the observers, including David Barrett, were still rooting for Flaco to be returned to the safety of the zoo. Then on February 9 Flaco climbed atop a cage with a live rat inside that had been set out by the rescue workers and briefly got his feet in the snare of netting meant to trap him, and once again the Flaco saga almost ended before it really got going. The early days would be full of these close calls.

And then it happened.

It might not have made headlines in the *Times*, but it did in the world of Flaco. Owls are elegant birds in many ways, but not in their manners at the dinner table. They gorge and choke their prey down whole—quickly and aggressively. Anything digestible continues on its journey down their digestive tract, but anything that is not—fur, bones, teeth—gets spit back out in a compact ball called a pellet.

And that was the big news, the cause for great excitement within Team Flaco in mid-February. Flaco had expelled a pellet!

Which meant he had eaten, which meant he had hunted. The thrilling news moved quickly among those in the growing birding and online community.

THINGS WERE HAPPENING FAST in Flacoland. "He became the world's most famous bird in a matter of days," David Barrett says. What David doesn't say is that at the same time he was becoming the world's most famous birder.

"My followers began posting about Flaco right away," he told me during our first phone call.

Follower is a funny word. It was what David always called those who had joined and posted on @BirdCentralPark. This makes perfect sense, since that is what X itself, and Twitter before it, designated them, but there was something slightly regal about the way he said it. He would also sometimes refer to one of his "famous posts."

In some ways David was an unlikely bird expert. He had come to birding relatively late, though he made up for what he lacked in experience in drive and intensity of focus. He had approached birding through a kind of self-created and self-willed program that reminded me of the Charles Atlas ads where readers were challenged to "make a man" of themselves. In a similar manner David Barrett had made himself a birder.

David's origins were murky, intentionally so it seemed. He liked to say that he had rarely lived away from Manhattan except for his education in Boston and Chicago, but he never mentioned his actual birthplace. As an undergrad at Harvard he was that creature known as a math major, and numbers and statistics would guide him through his life. After graduating from college in 1986 he continued to study math as a doctoral candidate at MIT, thinking he might become a math professor, but then he changed course, getting his MBA at the University of Chicago and returning to New York and Wall Street, where all that math could be spun into gold. Along the way to becoming Central Park's go-to birder he started his own trading company, studied classical singing, and trained as a runner in the mile event, hitting times in the low fives, exceptional in his age group. Somewhere along the way he picked up what most people would consider a British accent. When Nell Porter-Brown, a reporter for *Harvard* magazine, asked about the accent, he told them it was an "RP accent. RP British . . . received pronunciation." Since I didn't know what that meant, I looked it up and found that it means "the standard form of British English pronunciation, based on educated speech in southern England." This didn't really help solve the mystery since he also told the magazine he had only visited England once and briefly.

Perhaps more tellingly, he told Porter-Brown that though he lived a spartan life and watched little TV, *Sherlock* was one of his favorite shows. This made perfect sense. As Porter-Brown wrote: "Like the great detective, Barrett is exceptionally rational and ultra-focused." He brought this approach to everything, including his running, where he meticulously charted his heart rate and times. After returning from Chicago, he almost never left Manhattan, eventually buying an apartment on the Upper East Side, and most of his training for racing occurred in nearby Central Park. It was running, in a roundabout way, that would lead him to birding. Once David hit his forties he began to compete in masters division races, and after leaving the firm he helped found and deciding to trade his own money, he figured that his times would only get better since he no longer had to make the long commute that took up a good part of his day. But when his times didn't fall, when they in fact got worse, he concluded, in his scientific Holmesian way, that the commute had actually *helped* him. Why? Because it had forced him to walk to the subway, walk to the train, and then walk to the office, and, it turned out, walking held a secret. As he wrote in *A Big Manhattan Year*, his self-published book about how he became a birder, "I was missing out on a lot of walking and—crucial to the distance runner—the mitochondrial development that comes with it." Walking would allow him to put extra miles in without the microscopic muscle tears that overrunning could cause.

The problem was that walking was boring. He was a creature of goals, and running through the park fulfilled those goals, but walking? Dull. His mind was the sort—math problems, classical singing, running—that needs constant goals. And so as he walked that mind wandered, and as it wandered he began to notice people who were noticing birds. At the time he knew nothing about birds—well, maybe the basic birds most people can recognize, robins and blue jays and pigeons and gulls and geese. But he decided watching them might be a good way to pass the time, a puzzle to solve while getting his mileage in. He approached this the way he approached everything, by researching, and must have felt a little thrill when he learned birders kept "life lists," records of every bird they had seen. He liked lists. He took his first official birding walk on November 28, 2010. He wrote of that Sunday stroll: "I had already picked up on the theme of list-keeping in birding, which was consistent with what I was doing to track my running progress, so I noted what I saw and put it on a spreadsheet afterward: mallard, northern shoveler, Canada goose, and house sparrow." Soon he was far beyond sparrows and geese. He pursued his new hobby assiduously and his progress was remarkable: "I had always sought out activities in which I could measure my progress objectively. Mathematics with the problem-solving contests I enjoyed as a youth was like that; so were my sports, such as golf and running; and so was securities trading. Being able to quantify my birding progress motivated me to continue learning more about the subject. I would study photographs and descriptions of species I had not yet seen to prepare myself to someday see them."

David had always been a good student, and he studied through that first winter, creating a "Spring Birds" document. When the warm weather and the warblers returned, he was ready. By April 12 his life list had reached sixty-one, impressive for someone who had only started birding four and a half months before, but not enough for him: "I should have been getting close to that many in a single day." Online postings gave him clues and revealed sightings like the elusive Louisiana water thrush or the green heron down at the Turtle Pond, and he marched right out of his apartment and checked them off the list.

That list was growing. And so were David's never-quite-latent competitive instincts. He wanted to see all of the two-hundred-plus species that have been known to visit Central Park. And he wanted to prove his chops as a birder. He would do so, and quickly. A "Big Year" is when birders compete to see as many birds as they can. Big Years can be confined to a specific place, like say, Manhattan. David launched his first Big Year in 2011, but it was really just a warm-up for 2012. That year he saw and noted 208 species, placing second to a seasoned birder and well-respected researcher named Andrew Farnsworth. By 2018 he had become the unofficial king of Manhattan birding, and his record for that year was 230. (David upped this record to 237 species in 2020 before deciding to give up "competitive birding" to focus on running his Twitter account.)

David would bring this same focus and intensity to his pursuit and chronicling of Flaco's days. He had set up his birding site, Manhattan Bird Alert, in 2013, and it had begun modestly enough, mostly used by hard-core birders. But by the time of Flaco's death it would have ninety thousand followers, and through his posts thousands of people would end up investing in the adventures of a certain eagle-owl.

HUMAN BEINGS HAVE LONG been drawn to owls. It isn't hard to see why. The round eyes, the haunting hoots, the large forward-facing head. Not surprisingly owls were with us from our artistic beginnings. In *The Hidden Lives of Owls*,

Leigh Calvez writes: "In Europe, hunter-gatherer people of the Stone Age carved a Long-eared Owl in the Chauvet Cave, home of the second-oldest cave paintings in France at about 32,400 years old."

There is a mythic, storybook quality to owls, and that is true especially for eagle-owls. Jennifer Ackerman, one of our very best writers on birds, writes: "The name of the Eurasian Eagle Owl evokes a hybrid mythical creature, half eagle, half owl, and that's not too far from the truth."

Later, in her book *What an Owl Knows*, Ackerman speaks with Robyn Fleming, a research librarian at the Metropolitan Museum of Art, who described the history of owls in art from Goya to Picasso (who come to think it, had some pretty owly eyes). Of particular interest to Flaco fans might be the "huge Russian woodcut of a Eurasian Eagle Owl" that Fleming discovered in storage in the basement of the museum.

This is the artistic tradition that David Lei, a birder and photographer, worked within during his year of chronicling Flaco. He would become one of the most steadfast and inspired of Flaco followers, producing hundreds of beautiful photos and dozens of videos of the owl.

David Lei was also one of the best Flaco observers. Recently, when I scoured through the thousands of tweets and responses from the Flaco year, I found that his posts stand out for their general reasonableness of tone, a calm that doesn't fall far short of wisdom.

David Lei also happened to be one of the first to see Flaco hunt. On February 13, less than two weeks after Flaco escaped, he posted: "I first saw Flaco, the escaped Eurasian eagle-owl, hunt and eat Thursday night. Waited to share this so that it could get to the zoo staff first and allow them to adjust as needed. Flaco has been observed hunting each night since!"

THE PREVIOUS THURSDAY WOULD have been the 9th, the same day Flaco had almost been caught in the trap, which meant in just one week the former zoo-dweller had learned how to hunt on his own.

Almost as soon as it was clear that Flaco could hunt for himself the online narrative began to shift, and the Free Flaco movement was born. At the same time many ornithologists and expert birders warned of the dangers, both to and potentially *from* Flaco. By March 9 an online article in *Audubon* magazine gently scolded the freedom crew, stating unequivocally that the bird was in danger from a variety of urban threats, including rat poison, and was also a threat to native birds. It belonged back in its cage. But others noted that the cage itself was much smaller than that recommended for the species and that the only animal the owl seemed to be threatening was New York City rats. Things heated up on X, with plenty of attacks, attacks on the Flaco-watchers for crowding the bird, attacks on the zoo for persisting with trying to catch him, and attacks on David Barrett, made a target by virtue of the prominence of Manhattan Bird Alert.

One of the main criticisms was that by popularizing the owl and posting his whereabouts Barrett was responsible for the crowds that stared up at the bird, getting in the way of the efforts to capture Flaco and potentially bothering him.

Jim Breheny, the director of the Bronx Zoo, said so directly while retweeting one of David Barret's posts: "Whatever your intent, your need to seem relevant or involved in this effort is not at all helpful. In fact, you are a hindrance."

Some of this criticism was reasonable. With dozens of people gathered around trying to take pictures, the efforts to trap the owl were certainly more challenging, and by giving out Flaco's exact location, David was bringing more people to the owl. But at that point the genie was out of the bottle. There was no stopping the Flaco momentum, and even if Manhattan Bird Alert had suddenly shut down, someone was always going to be revealing the famous owl's location. "One lesson of the internet age is that information will get out," David had written in his birding memoir a decade before Flaco came on the scene.

As for Flaco himself, when it came to crowds or people staring at him, he was an old pro, pretty unflappable, pun intended. Also, the crowd tended to self-police and most of the observers were on good behavior.

Reasonable discourse, however, is not what Twitter/X is known for. Online trolls came at David Barrett like the mobbing crows that harassed Flaco. Extreme members of both sides villainized the other. The freedom lovers were accused of being irresponsible soft thinkers—"cat-loving sentimental types" was the term I heard, long before vice presidential candidate J. D. Vance came along with his talk of "childless cat ladies"—who didn't understand that a domesticated owl was in no shape to deal with the multitude of threats, both animal and urban. Meanwhile the Wilderness Conservation Society and the zoo were portrayed as villains out of Dickens, stifling Flaco's treasured freedom.

David Lei proved a calming influence as voices rose. He tweeted:

> At the beginning of the weekend, there was elevated concern about Flaco. Reasonable concern from knowledgeable folks, false outrage from people seeking to generate likes and everything in between.
>
> I think the simplest explanations for why Flaco hasn't been rescued yet are the most likely and why the zoo is pausing rescue efforts: 1) He is in good health and adapted quickly to survival in the wild; and 2) he's developed a taste for giant Central Park rats.

Then on February 16 he posted a beautiful photo of Flaco on a rock outcropping with this description:

> Flaco on a rock-o. Our escaped Eurasian eagle-owl has used Central Park's rock formations in a way that his wild relatives overseas might use the rock outcroppings and cliffs of their preferred habitats. Lights of the Upper East Side in the background.

(A perfectly reasonable observation it seemed to me, but alas David had used the possessive at the beginning of the second sentence, and one respondent scolded: "He's not 'yours.'")

Then on February 22 he posted again:

> Flaco, the escaped Eurasian eagle-owl, perched on the Naumburg Bandshell last Friday night. Extreme opin-

ions have been expressed about him. On this subject and others, perhaps we should behave more like owls: patiently watch and wait.

MEANWHILE, SIMULTANEOUS WITH THE development of the *Keep Flaco free* movement, another narrative was growing: *Flaco is famous!*

Within days of his release Flaco was becoming known throughout the world, and within a couple of weeks Flacomania was cresting. In a society that seems to value fame above all else, Flaco had it. And it was rubbing off on David Barrett. He was the go-to quote guy in the *New York Times*, *Time*, and the *New Yorker*. The networks all covered Flaco as did *Good Morning America*. The *New Yorker* would eventually publish a nice little absurdist piece by Evan Allgood, written in the voice of another bird who visits Flaco and finds he has gotten too big for his britches. Flaco brags about being on the cover of the *Post* again, and the bird narrator notices he has "dropped his Carolina drawl and affected an old-school New York accent, like Ryan Gosling." This Flaco likes to pose, hoping David Lei is nearby to take his picture, and he sometimes visits the Beresford so he can loom "atop a tower, watching broodily over the city like Batman."

Manhattan Bird Alert's popularity soared. Just before Flaco escaped, David Barrett's account had 65,000 followers. It would grow to 91,000 by late February 2024. A post about Flaco returning to the park after flying to the Lower East Side got 316,000 views and 6,200 likes.

Later, when Flaco departed the park in November, the post trying to explain why he left garnered 556,000 views and 2,500 likes. And by the end of the year, David's report of the bird's death received 1.4 million views and more than 9,900 likes. Maybe not Taylor Swift numbers, but within the birding community this was *big*.

When I asked David Barrett if it was fair to say that *millions* of people followed Flaco he replied: "Flaco's story was featured prominently in the *New York Times* (often), on network news, and in the major news sites of many foreign countries. The *New York Times* alone has a circulation of over 10 million. It seems likely that millions of people were interested in Flaco's story and that perhaps 50 million people or more had some awareness of it."

Old-school birders steamed, irked by the attention that David Barrett received. Why ask *him*? He wasn't even a proper birder, let alone an ornithologist. This wasn't the way it was done. What they didn't understand was that it was a new world we were all living in. David Barrett had figured this out before anyone else.

Flaco was not the park's first celebrity bird, not by a long shot. He was really just the latest in what had become a proud New York tradition. Two decades earlier Pale Male, a red-tailed hawk, had been the star of Marie Winn's bestseller *Red-Tails in Love*. Not long before Flaco in 2021, it was Barry the barred owl who ruled the park roost and became the cynosure for park birders. The extroverted two-year-old owl was the darling of the city. A year later Rover the bald eagle showed up at the park reservoir and entertained watchers by taking out gulls in midair, feathers flying everywhere. Rover would make a triumphant return in January 2024, briefly sharing the spotlight with Flaco himself. There had been other stars, like the mandarin duck, who birders named Mandarin Patinkin or the Hot Duck, a rarity from East Asia (having likely escaped from captivity, not flown across the ocean), who became both a movie star (of a documentary called *Birders: The Central Park Effect*) and the subject of a children's book by Bette Midler. In fact, it was David Barrett's posts about the Hot Duck that had marked the beginning of the rise of Manhattan Bird Alert. "I've had a few celebrity birds—the mandarin duck was my first," David told the *New Yorker*.

Perhaps more interesting, to me at least, was the snowy owl who showed up in the park in January 2021. This was the first snowy seen in the park since 1890, a 130-year gap. In recent years there have been several snowy "irruptions," winters when these birds, who usually make their homes in the Arctic, appear in large numbers in the States. This often occurs after there has been a boom in the lemming population in the Arctic, which in turn leads to a boom in the owl population, pushing young birds south in search of new territory. I was lucky enough to have been on Cape Cod during the great irruption that marked the winter of 2013–14, when several snowys also made their way down to New York. The owls made a great splash, in both social and regular media, a Beatlemania for birds. But the Beatlemania metaphor only works to a point. When the Fab Four landed at JFK to launch their first U.S. tour—almost fifty years before the owls to the very day—they were rushed by frenzied teenage girls. When the snowy owls landed at the same airport, they were greeted by men with guns, hired by the Port Authority, who actually shot and killed three of them, worried they would interfere with air traffic. (Airports, like beaches and marshes, have that low, flat, treeless look that to the snowy mind says tundra.)

Most of these feel-good urban wilderness stories have not-so-happy endings. Barry would swoop down from a tree and hit a maintenance vehicle driven by two park workers, and Rover would collide with a truck, or rather a truck would collide with Rover, not long after his triumphant return in February 2024. Auto collisions had also killed Rover's mother and one of his siblings. Meanwhile Barry's necropsy revealed potentially lethal amounts of rodenticide, rat poison put out by the city. Even Pale Male wasn't immune, having perhaps unintentionally poisoned his young by bringing back poisoned rats to the nest. Of course it isn't just celebrity birds who suffer these fates. Red-tails and other raptors without cute names die all the time from rat poison and collisions, both with trucks and with buildings.

From the beginning there were warnings mixed in with the celebrations of Flaco's freedom, and most everyone was aware of where this all might be heading. We've all seen enough movies about stars who make it big but who meet with tragic endings. Jim Morrison. James Dean. Janis Joplin. Kurt Cobain. Flaco?

It didn't take a whole lot of wishful forgetfulness, however, to focus on rooting for Flaco as his life in the park expanded. Worries about rat poison and other threats were more than counterbalanced by excitement about the chance that Flaco might just be able to make it outside his cage. Soon, as he began not just to hunt but to fly and hoot, more and more people began to follow him on his adventures, and it was easy enough to block out the darker tidings.

3

EVOLUTION

BY MID-FEBRUARY, two weeks after Flaco's escape, with public sentiment welling behind keeping Flaco free, the Wildlife Conservation Society, and therefore the Central Park Zoo, started to waver in its commitment to capturing the owl. Whether or not this was in reaction to the growing Free Flaco movement is not entirely certain, but public opinion sure seemed to impact how they proceeded.

The Wildlife Conservation Society, which would become an online whipping boy over the next twelve months, had noble origins. Founded as the New York Zoological Society, it came into being when Teddy Roosevelt, then president of the Boone and Crockett Club, founded the society with three stated objectives: to open a zoological park, to promote the study of zoology, and to preserve wildlife. Modern zoos, combating the idea that zoos are still just menageries of captive creatures, are quick to point out that part of their missions, aside from entertaining and educating the public, are the crucial work of field conservation and species recovery, and the New York Zoological Society provided both the earliest and most famous example of this when they reintroduced fifteen American bison, a species on the brink of extinction, to what was then called the Wichita Forest and Game Preserve.

The WCS continues to work toward species recovery in a time of massive biodiversity loss and has a stated goal of trying to help save 50 percent of our remaining wild lands. All fine and good, but when it came to Flaco, they were caught in a public relations bind. During the first two weeks after his escape, the organization was intent on recapturing the owl, and came close a few times. But once Flaco caught and ate his first rats, public opinion, fueled in part by Manhattan Bird Alert and articles in the *Times*, got behind the idea of a free Flaco. At that point the WCS softened their stance, saying they would "continue to monitor him, though not as intensely, and look to opportunistically recover him when the situation is right." But then, only five days later, they set a baited trap in the Sheep Meadow and tried to lure in Flaco using a recording of a hooting female owl. "That was a bizarre night," remembers David Barrett, who was out following Flaco. Online outrage followed—"The nerve of them, the deceit!"—and battle lines were quickly drawn. David Barrett posted that there was a rumor an actual female eagle-owl might be used to lure Flaco in. Others posted that the crowds David was drawing in were getting in the way of capturing Flaco and therefore threatening the bird's safety.

"It seemed like in those early days, the zoo was going back and forth a little bit," David remembers. "So initially the zoo called off its recapture efforts, but only temporarily and mainly because they didn't have the right equipment, I believe. And Flaco seemed to be alight. There was no overwhelming need to capture him. Roughly a week after his release, it was known that he could survive on his own."

In retrospect, it isn't hard to empathize with the bind the WCS and zoo were in. Raptor experts have suggested to me that Flaco could have been captured successfully with a cannon net. But a device that used explosives to launch a net was a potential public relations nightmare. The zoo workers were well aware that they were onstage while they went about their business, which is why they did a lot of it at night. "What happens if they injure or, God forbid, kill Flaco while attempting to capture him?" Richard Farnsworth asked. "It adds to the whole notion of the whole thing's onstage. And the capturers are onstage too."

The world was watching.

IMAGINE FLACO FLYING THROUGH Central Park as darkness descends.

Owls are known as the lords of the night, but it is during the crepuscular hours that eagle-owls begin their reign. Flaco would have flown silently, his wings holding an evolutionary secret humans had long noted but had only recently begun to understand. Those wings are made up of feathers that, unlike those of other birds, reduce turbulence and the noise it brings and are coated with pinnulae, fibers that muffle sound. When Canada geese or even swans fly overhead you hear the loud whoosh and flap of their wings. Not so owls (though if you are close enough, the whoosh of displaced air can't be disguised).

Though everything was new in the park, Flaco's eyes—enormous, round, forward-facing like a human's but making up a staggering 5 percent of his body weight—were another great advantage. This was thanks to a layer of tissue called the tapetum lucidum, a retroreflector, a kind of internal mirror that reflects visible light back through the retina—and to the fact that those eyes are packed full of retinal cones.

While Flaco's eyes might have been his most obvious and arresting feature, it is the organs they worked in concert with, his ears, that were even more subtly dramatic. Not the flamboyant tufts that give the bird a horned look but don't actually aid in hearing, but the ears themselves. Owl ears are slightly asymmetrical, with sound reaching one a millisecond or two sooner than the other, allowing the birds to home in on their prey's exact location and, working in conjunction with those magnificent eyes, find that prey in near total darkness.

These were the tools that evolution had gifted Flaco.

The problem was that he had not yet learned how to fully use them.

"It was like owning a Maserati but not knowing how to drive," was how one Central Park birder put it.

Being hand-fed and unable to fly more than a few yards from side to side for a dozen years had stripped Flaco of his instinctive abilities.

What was fascinating for Flaco-watchers in those early weeks after his escape was the story of growth, of evolution, of watching him regain those native skills.

IN EARLY MARCH, A month after Flaco escaped, I happened to be in town, far from my home territory. I was there because my daughter, Hadley, was a freshman at New York University, having, like Flaco before her, moved from North Carolina to New York. On the morning of March 9 my wife, Nina, and I took the subway up from Union Square to Central Park and spent the better part of the day searching for Flaco. Hadley had not been caught up in the early waves of Flaco-mania and was amused we were going to spend our day in the city owl-hunting. The fact that she had chosen New York as her college destination had come as something of a surprise, if not a jolt, to her parents, who had imagined her frolicking on the grassy lawns of, say, Middlebury. But she had decided she wanted to make films and also decided, completely on her own, that the Tisch School of the Arts at NYU was the place to do it. I had studied ospreys some years before and noted that the process of fledging chicks is, no matter how helpful and intense the parenting, ultimately abrupt, and our only child's departure from our home after nineteen years had been unsettling, especially for Nina.

We spent the day exploring the park, focusing on the hills and hollows of the North Woods, where Flaco had recently been seen and where he had one of his favorite roosting trees. In finding the North Woods it was Flaco's good fortune that the territory he'd discovered was remarkably like that which Eurasian eagle-owls had evolved in over the eons, a forest of trees and rock ledges and water in the form of lakes and creeks and a small waterfall similar to the Eurasian forests that had helped make him who he was. True, the North Wood's dimensions were not quite as vast as those where his ancestors had evolved—0.3 miles wide and 0.45 miles long—but in a city of 7.93 million people, you take what you can get.

Though the calendar claimed it was still winter, it was a beautiful spring day. The cherry trees were bursting with pink and meaty buds as we walked through the park, past the Pool and through the Glen Span Arch into the small patch of boreal forest where the world's most famous owl had been spending many of his nights. Flaco had been free for just over a month. We stopped by the small waterfall where green moss covered the rocks and a great gnarled Osage orange tree cantilevered out over the water, a massive broken branch hanging down and not quite touching the surface, like a man's wounded arm. Flocks of field sparrows and chickadees and titmice gathered in the briars, tame and apparently used to handouts from park-goers. Black locusts and hickories towered over the water.

Unbeknownst to us, we passed right by the tree that would become Flaco's favorite, a black oak that grew straight for twenty feet before spreading its branches out like an open hand. A year later I would return to that tree, only two days after the Flaco memorial, and find it surrounded by flowers: garden mums, Peruvian lilies, wild daffodils, tulips, carnations, and baby's breath. I had watched the memorial online, and though I found it a tad New Agey for my taste, it was moving that so many people were affected by the death of an animal.

We spent a few hours exploring the woods and the nearby areas, including the ballfields of the North Meadow and Harlem Meer where Flaco had been known to rest atop construction equipment. Since Hadley had moved to New York, it always surprised me what a workout our days there ended up being, and I would sometimes look down at my phone in the evening and see I had walked sixteen miles. I was in decent shape, training for a western hike on the Arizona Trail, and at one point I half-ran up the Great Hill. I was caught up in the excitement of the chase, sure we would find our bird, who was likely roosting and camouflaged as he slept away the day.

The closest we came was a bulky shape in a tree, half-hidden, and I called for Nina to come join me atop the Great Hill. I watched it carefully, growing more certain its shape said owl, until at last it flew off and revealed itself to be an immature red-tailed hawk. We pressed onward, asking anyone we ran into if they had seen an owl, and, finally, breaking down and tweeting, hoping someone could give us the owl's coordinates. (It was still Twitter then, becoming X in the early hours of July 23, 2023, about halfway through the year Flaco spent outside the zoo.)

If I'm a fairly hapless birder—naturally impatient with a tin ear—I'm a completely hapless tweeter. I have since deactivated the account I was using that day and so can't recover the day's tweets, nor can I recall the exact advice I got as we tried to chase down the bird. We certainly could have used David Barrett's help that day, though for all I know it was David himself who tweeted back and told us that

Flaco had disappeared and that there had not been a sighting of him that day. Of course I would have loved to have seen him, but there was something satisfying about the day despite our failure to achieve our goal. It felt right. To be outside tracking an animal. Millions of years of evolution would have made Flaco feel comfortable in the North Woods. Those same millions of years had done something similar to us.

During the year following our March search, Nina, along with thousands of other people around the world, would continue to track Flaco online. I would not. She would report to me what Flaco was up to, but I only rarely frequented Twitter/X. It wouldn't be until almost a year later, with Flaco recently dead, that I would resume that day's explorations, and when I did so it would be on foot, not online.

Incidentally, Flaco wasn't the only New Yorker who Nina tracked. Thanks to her phone app, Life360, she could follow Hadley through her days, looking on as our daughter headed from her dorm on Fifth Avenue over to the Tisch building for class, and, more intently, through her nights, as she headed off to parties in the East Village and Brooklyn. Like Flaco, Hadley was primarily nocturnal, and like the owl, she was learning her new territory, becoming adept at using the subway and avoiding the murderous electric bikes that screamed down the bike lanes, sometimes going the wrong way. Like many a New Yorker before her, she assumed an attitude of amused condescension toward out-of-towners like her rube parents, visiting from the backwater of North Carolina.

There was a tension, at least for us, between Hadley's safety and her freedom. She reveled in having the city as her new territory, and we were happy for her. But we were also scared.

BY THE TIME WE were hiking through the park, I knew Flaco could catch his own prey. Naturally enough I imagined an eagle-owl hunting in the wild. In my mind's eye I saw him sitting on a tree branch as dusk descends, bobbing his head so that his satellite dish of a face could help direct sound toward those acute ears, peering out with those blazing orange eyes. And then he hears it, a faint rustling perhaps, a deer mouse under the leaf litter. His ears and eyes, working together, pinpoint the mouse's exact location. At that moment the great bird drops silently from the tree with deep, slow wingbeats, powerful flaps with his wings up above his body. He drops lower, gliding toward the poor animal, diving down headfirst before kicking back with his powerful talons, crushing in with a violent burst. The mouse hears nothing until it is too late.

But that was not how Flaco actually hunted, at least not at first. During those early days he did swoop for prey on the ground a few times, but unsuccessfully, and during his first known successful hunt, he caught a brown rat by chasing after it on the ground. The chase was more comic than grand. In fact hunting on foot remained his m.o. for most of those first few months.

But if his hunting remained rudimentary early on, his flying was improving. He was no longer exhausted after short bursts but was gaining stamina. By late February he had, according to David Lei, "expanded his range and explored far to the north, even going undetected some days, and his flight skills were improving dramatically."

Over at the Manhattan Bird Alert, David Barrett was singing the same song. He later summarized his thoughts for a publication called the *West Side Rag*: "Not only has he developed the skill and endurance to fly around the park, he looks graceful doing it. This after a lifetime of being forced to be mostly sedentary. He hoots, too! Probably both to indicate his territory and to find a mate."

Barret continued, "After spending a couple days last week resting over his old home, the Central Park Zoo, he has moved on to explore the park. He seems to be enjoying his freedom, doing all the things a wild owl would be expected to do."

LIKE A LOT OF good nature books, Henry David Thoreau's *Walden* included, Flaco's story would be contained in the story of a year. Henry David Thoreau's stay at Walden Pond had actually lasted more than two years, which he had shaped into a single year for the sake of his book. Flaco's story did not need to be so fudged, as his time outside the zoo would last almost exactly a year, from February to February. In nature books the story of moving through a year is often one of phenology, of leaves budding and later falling, of young birds being born and migrating, of life and death. That was Flaco's story too.

Thanks to Thoreau's writing, both his famous book and the journals he so regularly kept, you can go back and follow him through the "year" at Walden. And thanks to the many online chronicles and websites and Twitter/X accounts, you can do the same with Flaco. You can watch him during those first days as his territory expands beyond Hallett, and as he returns to the zoo for a visit, and then as he flies to the Great Lawn. You can tune in when the action picks up on Day 8 when he is almost caught in the trap laid out on the ballfield lawn, scoping out that white rat inside the wire cage put out by the zoo, taking his time before jumping on top of the trap, getting his foot entangled in the fishing line meant to snare him, but shaking free and flying off. And you can watch him the next day while he takes his time devouring a rat he has caught all on his own, sitting up on the branch of a white pine, the rat's tail dangling down like the last strand of spaghetti.

You can continue following him on the evening of February 12, the night of the Super Bowl with the park eerily quiet as people hunker down in front of their TVs to watch the Chiefs play—who else?—the Eagles. By then he is hooting strong, in a long bout of a hundred hoots, and another of eighty. That will be the night of a couple firsts: the first night he is seen perching on a backstop of the Heckscher Ballfields and the first night he begins to explore a construction site. A couple days later on Valentine's Day the online chatter turns to thoughts of romance when Flaco edges into the territory of Geraldine, the park's resident great horned owl. The two species have actually interbred in captivity, which leads to both romantic effusions and stern warnings about Flaco interfering with the native wildlife, though no one actually witnesses the two interacting.

You can then experience Central Park as winter gradually edges into spring and Flaco gracefully endures his one snowstorm while the trees start to bud. The videos reveal not just sights but sounds. Flaco had escaped the dreaded song of the Delacorte Clock, with its mallet-wielding monkeys, but he hadn't escaped the noises of an overspilling city, and trucks and horns and sirens and fireworks were the soundtrack of his spring and summer in the urban wild. In some of the films of Flaco you can hear trombones and kids playing and what sounds like a brass band and always, always, there was a noise he must have grown used to: the clicking whirring electric typewriter sound of cameras taking thousands of photos. In his own way Flaco was now adding to the symphony, hooting more regularly, haunting hollow *whoo*s that his followers, and anyone else within earshot, couldn't seem to get enough of.

There are surprises along the way. About a month in, Flaco is sitting peacefully in his tree when he is broadsided by a red-tailed hawk. He takes the hit and flies off his branch but is none the worse for it. He also endures encounters with Cooper's hawks and other red-tails and blue jays and tufted titmice, as well as taking a minor body blow from an aggressive squirrel.

Online viewers got hooked on the plot. "There was an adventure element to the whole thing," said Dave Propson, who I first met on X as @bangkokdave. "Flaco and a crow. Flaco meets a squirrel. And suddenly you were sucked into the story. The adventure."

The adventure grew more exciting as the trees leafed out and Flaco began making passes at his prey in flight, eventually starting to catch the occasional brown rat on the fly. He would still pounce on rats and eat them right on the ground, but by late May he would sometimes drop down from a tree, "swooping for a rat, catching it, killing it in his talons and continuing on to another perch without landing," according to David Lei. By summer Flaco-watchers were reporting that he often ate as many as four rats per night.

He developed his skills the way most of us do. Slowly, by accretion, but then also suddenly jumping forward. As his skills expanded so did his horizons. The North Woods, the Ramble, the Sheep Meadow, the Loch, the Harlem Meer. Darker stories would later be told in the wake of the grim necropsy report, with at least some of the early warnings proving true. But what we forget when we render these retrospective judgments is the sheer thrill of those first days. They were days of growth, of change, of evolution. And while it may be dangerous to attribute specific emotions to an animal, it is fair to say that, at least on a physical level, a kind of self-realization was occurring. Certainly Flaco was becoming more. Parts of him that had long slumbered were waking up.

Would it be going too far to say that as he got stronger Flaco took some pleasure in flight? Maybe. But in *When Elephants Weep*, Jeffrey Masson speculates that animals feel *funktionslust*, a German word meaning the "pleasure taken in what one can do best." It must have been a release to fly without restriction, to travel more than the length of a department store display window through the air. Many commentators would use the word *freedom* to describe Flaco's experience, but as recent political trends have taught us, *freedom* is a slippery word. Space is less abstract, and space is what Flaco now had.

"I love a broad margin to my life," Thoreau said.

Perhaps owls do, too.

DURING THOSE MONTHS OF growth, it wasn't just the owl who was changing. Those who wanted to follow him during his active hours, and not just look up at him as he snoozed away the day, were becoming by necessity nocturnal hunters. They adjusted their schedules, becoming day sleepers like the creature they were following. Excitement helped them cut down on sleep altogether. A core group of photographers, four or five some nights, fewer other nights, would sometimes stay until the park closed at one.

By summer Flaco was seen sipping water from the tennis courts and exploring the park's northernmost reaches, enduring more battles with Cooper's hawks and red-tailed hawks, but generally making the larger park his home. He found a meadow—the Central Park North Meadow officially but "his meadow" to his followers—that was locked from dusk to soon-after-dawn and made for fine hunting, which varied his diet somewhat, adding white-footed mice to his staple of rats. Over the summer his hooting died down, but it resumed in mid-September. With fall he began to feel a new restlessness.

By then Flaco had become both a symbol and an individual that people had gotten to know, a powerful combination. Flaco-watchers had grown so used to regularly seeing the owl that they could imagine he would live out his life in the park. But as it turned out he would not.

Exploring the park, his new territory, would be just the beginning. The park was not an enclosure, and there was no steel mesh to keep him in. There was a world of concrete and skyscrapers to explore outside its borders, and over the next months that, too, would become part of Flaco's territory.

4

VISITATION

FOR EIGHT MONTHS Flaco made the 843 acres of Central Park his home; for eight months his followers tracked his every move. Though much of the park is forested and good owl habitat, Flaco's hunting grounds were some of the most urban spots in the park, including the compost pile and construction site near the North Woods and the larger construction site where the new public pool/rink was being built near the Harlem Meer. There he liked to perch not atop craggy rocks but heavy machinery, and he was often seen atop cranes (the machines not the birds) and tractor blades and bulldozers. Why did he favor these less aesthetically pleasing spots?

Rats of course. Big, fat, appetizing New York rats.

The dug-up dirt of his semiurban hunting grounds were full of them.

The photos of Flaco, and they were legion, often showed him looking contemplative or cute or wise (or all three).

But they also showed him eating, the red blood of his prey almost as vivid as the orange of his eyes. Back when I studied ospreys, I noted the combination of grace and savagery the raptors exhibited, never more than when they fed their young in the nests. They would savagely tear into a fish, ripping off a chunk of flesh, and then oh-so-delicately place it in the gaping mouths of their nestlings. Flaco, without any offspring to bother with, could do away with the graceful part. He tore into the rats, manners be damned, and wolfed, or owled, them down.

But while he had food, another urge went unfulfilled. European eagle-owls were long thought to mate for life, though lately their loyalty has been downgraded to "mostly monogamous," a designation that also applies to ospreys and many human beings. Flaco, however, lived in a kind of enforced celibacy, first in the zoo and later in the park and city. Every night he hooted and never heard a reply, his calls for a mate going unanswered. He never did start up a romance with Geraldine, and his other options were limited. In all likelihood this unfulfilled desire for sex, for procreation, played a major role in pushing Flaco to leave the relative safety of the park. After months of fruitless hooting, he was going in search of a mate. Of course, what he would learn was that he was the only living eagle-owl in New York. There were no mates to be found.

Whatever the motivation, on the night of October 31, Flaco suddenly left the park and headed downtown.

David Barrett posted that night: "Spooky and scary Flaco opens wide to try to expel a pellet, hoots, stretches, and hoots some more from his favorite oak tree in Central Park. Happy Halloween!"

But that was it. Over the next six days no one saw the owl. With each passing day Flaco Nation grew more nervous.

Then, at the end of the first week of November, David got a tip that Flaco had been seen in an enclosed green space near the East Village. After all those months in Central Park, Flaco had flown almost six miles south.

FLACO WAS FOUND WELL hidden in a tree in the Kenkeleba House Garden on East 2nd Street near Avenue B. It is late April, two months after Flaco's death, when I visit the garden with Hadley. I'm guessing that at this point she might be tired of having her father fly up to New York every other week, but she dutifully leads me there from her dorm to the garden, eschewing my MapQuest and taking a new New Yorker's pride in knowing the way. Coming here from North Carolina, where the world is bursting green, is like taking a step two months back in time, but while elsewhere in the city the trees are still in early bloom, at Kenkeleba House Garden the season seems to have jumped ahead. What we find, locked behind a gray gate, is a lush and overgrown woods, only fifteen yards wide but three times as deep. It is part secret garden in the midst of the city and part Miss Havisham's, stone steps weaving through an overgrown forest of multiflora rose and shaggy trees covered with vines of oriental bittersweet and English ivy. Two free-standing gray statues that look just like pictograph figures I've seen on sandstone walls in southern Utah stand feeble guard over a lush and dissolute landscape. Hadley points to a rat scurrying through the underbrush, which, along with all the hidden spaces, suggests why Flaco favored the place.

On the walk back I get a text from David Barrett suggesting we stop at 60 First Avenue to see what he calls "the famous Blue Grosbeak." This is the sort of text Central Park birders are always sending, alerting each other to rarities. Hadley is dubious when I say we should walk into what appears to be an apartment building, but when we do and I begin to explain to the doorman why we are there, he cuts me off. He has clearly heard what I have to say before, and without another word he ushers us out the back door to a small garden behind the building. There we find two birders, heavily armed with binocs and cameras, pointing telephoto lenses at a rather lifeless and frumpy blue bird that sits in the dirt in the middle of a garden of China rose and tree peony.

When we shake hands and I make some small claim to being a birder myself, they are skeptical. The one with the bigger camera asks me why I don't have binoculars. I decide not to defend myself and instead ask another question: Why does the bird look so listless?

"He's probably sick," big camera says, then clicks another picture. Sick or not, a rarity is a rarity.

We stop and eat at the Vegan Grill where—you'll have to trust me—the fake scrambled eggs and sausage are almost as good as the real thing. The time of my flight is closing in, so I pay for the meal and kiss Hadley goodbye, leaving her eating her French toast. I make one last hurried stop, walking down 13th, a beautiful street lined with littleleaf linden trees. This was another stop on Flaco's East Side tour, only four blocks from Hadley's dorm. During his stay on the Lower East Side, Flaco began perching atop buildings, including co-ops and public housing, his hoots echoing through the streets. Down below, kids cheered on their new neighborhood mascot.

WITH FLACO FLYING THROUGH the city itself, his story took on a new attraction, the appeal of the urban wild. I have never lived in New York, but a spring I spent teaching in Cambridge opened my eyes to the flavor of this particular brand of wildness. Coyote tracks on the talcum of snow on the Charles River, red-tailed hawks nesting below the press box in Fenway, the nests of straw and sticks of a sparrow's nest drooling out of the mouths of gargoyles atop buildings. Hadley was born that year, May 7, 2003, just as the city was bursting with green after a cold and snowy winter. In the two years prior we had lived in a beautiful place, a house right on the beach on Cape Cod that we rented for the offseason, dog-sitting to reduce our rent. I referred to our move to the city as the beginning of "our time of captivity," but I was happily surprised by the wildness I found there. The spring before Hadley was born I spent some time tracking coyotes in greater Boston with a young biologist named Jon Way. We followed one particular coyote, whom he had named Maple, by radio collar as she made her way through the wilds of Saugus. Maple skulked through alleys and jogged down abandoned train tracks and would sometimes haplessly hide behind light poles when cars drove by. Her den was behind Weylu's, a Chinese restaurant that sat regally above Route 1, and she was known to prowl behind Burger Kings and used car lots at night.

If I was worried that moving to the city would put an end to my wild times, imagine how I felt about the impending birth of a child. But what surprised me was how wild that experience would turn out to be. From the first second I held my daughter I felt something cracking open inside me, and on her first afternoon, when I picked Hadley up while Nina slept and danced to the James Taylor tape that was playing, I found myself throbbing with tears. That it is a common experience does not make it a less profound one. But that it felt so wild, so animal, was what really surprised me.

Soon after Hadley's birth, our friend Karen, a doctor from San Francisco, said to Nina that you never feel so connected to the animal kingdom as when you have a child. Nina agreed. She had felt that way many times since the birth, but never so much as one morning when she couldn't get Hadley's face clean. Milk had crusted in our daughter's ears and above her eyebrows and the washcloth just wasn't working. So Nina did what came naturally: she put the washcloth aside and licked Hadley until she was clean.

What hit me daily with consistent impact during that time was the fact of my daughter's creatureliness. This squirming little apelike animal, barely two feet high, somehow had been allowed to live in the same house with us. Or as another friend of Nina's put it: "They're the best pet you'll ever have."

What I had been led to believe would be the most domestic of experiences turned out to be one of the wildest.

THOSE WHO WERE WELL informed about the fate of urban birds were not overly optimistic about Flaco's chances of making it in the big city. But Flaco continued to surprise.

The two Davids, Lei and Barrett, followed Flaco day-to-day as he explored his new territory:

> David Lei
> @davidlei
> Nov. 9, 2023
> I've spent the evening listening to Flaco's hoots echo between the co-ops and public housing buildings of the Lower East Side, not far from where I grew up in Chinatown. They remind me of hearing great horned owl hoots echo in the woods.

> Manhattan Bird Alert
> @BirdCentralPark
> Nov. 10, 2023
> Friday Flaco Update: Flaco was seen for the second consecutive evening perching and hooting atop an apartment building at the east end of Grand Street (near FDR Drive) on Manhattan's Lower East Side on Thursday from 5:30 pm to 7:00 pm.

> Manhattan Bird Alert
> @BirdCentralPark
> Nov. 11, 2023
> Here is Flaco the Eurasian eagle-owl in his Monday resting place in an East Village sculpture garden. He may find that similar places, gardens enclosed by buildings or fences, of-

> fer him peaceful sleep amidst a noisy and chaotic urban environment.
>
> David Lei
> @davidlei
> Nov. 12, 2023
> Flaco the Eurasian eagle-owl checking out someone's balcony in the Lower East Side of Manhattan Thursday night. Perhaps he heard that his cousins in Europe and Asia occasionally nest in balconies.

After that last post, on November 13, Flaco, having spent a couple of weeks on the Lower East Side, fell off his followers' radars again. As it turned out, he was making his way back north to Central Park, where his return home would be celebrated by his followers. On November 15 he would spend some time on another backyard air-conditioning unit of a Fifth Avenue apartment, and that evening David Barrett would see him fly back into the park.

But before he returned, he made one very important stop, one that is central to the Flaco legend.

IT IS RAINING HARD on the night of March 6 as I descend into the subway in Union Square and head uptown. My destination is the Fifth Avenue apartment of the playwright Nan Knighton. If David Barrett was the intentional leader of the Flaco year, Nan was the accidental star. She still relishes November 14, 2023, the afternoon of her visitation from the owl.

Nan Knighton knew nothing about the bird before it landed on the ledge outside her kitchen window. After that day she, like so many people in New York and around the world, became obsessed with the owl.

Nan is still in mourning when I visit her three weeks after Flaco's death. Mourning is not an exaggeration.

Nan and I texted yesterday and found that we had a lot in common. For one thing, we are both writers, and that is something of a relief. So far I have been interviewing mostly birders, who are articulate enough when it comes to talking about field markings but curiously inarticulate when it comes to what their encounters with birds might *mean*. When I bring up aspects of environmental philosophy or the meaning of wildness, they just stare at me for a minute before showing me a picture or mentioning a rarity they want to go find in the park.

I love birds, but I have always had some ambivalence, and perhaps a little reverse snobbery, toward birding. There were two reasons for this ambivalence. One, if birding is a sport, it is one I am not very good at. And two, the whole enterprise sometimes seems contrary to the spirit of wildness that drew me to birds in the first place. Over the years, however, my prejudice has slackened. I have spent a lot of time with expert birders and have become very impressed with their observational talents, their ears and eyes, basically what you might call their hunting skills. When I am with these people, I mostly keep my mouth shut, deferring to their expertise and talents.

Nan is not a birder either. She gives me a big hug when I come in the front door of her apartment, then immediately apologizes for the hug since we just met. But I tell her not to apologize. I get it. On the one hand we don't know each other. On the other hand . . . Flaco.

"I told my kids that other than their births the day Flaco visited might have been the greatest day of my life. That irritated them. 'What about our childhoods?' they asked."

She leads me into the apartment, which is exactly the sort of spectacular you might expect given its Fifth Avenue address, overlooking the park and reservoir. I'm happy to see the place overspills with books.

The kitchen where we are soon standing is where it happened. Right behind us at the window.

Nan stands at the counter.

"I was right here, and I'd just gotten off a phone call so I had my phone in my hand, and I looked up and there was an owl in the window. I thought I was hallucinating. An owl in the window! I saw mainly his head. He was on the ledge below the window, but he was a really tall bird. And I just started to laugh because it just seemed so wonderfully absurd. And because I love animals so much—I've heard some people say, 'Oh, it was scary'—but my reaction was, well, delight."

She laughs at the memory.

"And I didn't want him to fly away. We get red-tailed hawks that come to the front side of the building and land on the balustrades, and if you get near them, they fly away right away. I thought I should get a picture of this guy because he was going to fly away. So I gradually got closer, and he didn't fly away. He was letting me get closer and closer."

It was around noon, maybe closer to twelve thirty.

"I spent some time calling the Wild Bird Fund because I thought he might be stuck. I knew nothing about owls. I basically had never seen an owl close and so the whole thing was just crazy and wonderful, and I kept thinking, *He's gonna fly away, he's gonna fly away*, and so the whole time I get closer and closer and he's just looking at me and it was incredible."

Around the corner from the kitchen is a little nook with a desk that she used as a study, and from there she could get a full view of the owl.

"When I went into the office I would call back, 'Here, boy, here boy,' and his neck would swivel to see what I wanted. I know it was silly. As if he were a dog."

From the office you can also see more clearly that the window Flaco perched on looked out at a small courtyard, the kind he had begun to seek out to rest in during the day. Enclosed spaces like that allowed him to feel safe while he closed his eyes.

The owl stayed three hours. Nan would go back and forth from the office to the kitchen, just staring at the owl. Why? Because *Oh my god an owl on the thirteenth floor in New York City*.

Nan took five short videos of her encounter. The videos went viral.

"I wish I'd seen him fly off," she says.

I ONCE SPENT A few days paddling a canoe the length of the Charles River, weaving my way along the sinuous waterway into Boston. Traveling through backyards and past ancient abandoned factories and under highways and paddling past both stunning marshes and abandoned grocery carts, even stopping one evening to climb out of the canoe to drink a couple of vodka gimlets and eat a steak at the Irish Alehouse by Route 1 in Dedham, I had also heard coyotes at night and been ushered along by great blue herons. It wasn't the Yukon or even Yellowstone, and I knew I was heading into the heart of a city of almost five million people, but the trip held a wildness of its own. A limited wild, an urban wild, a backyard wild. It seemed promising to me that you could wander down behind a 7-Eleven and find a little wilderness of your own.

Wildness, like *freedom*, is a word that often gets waved around without much thought. What does it mean exactly? Is it just something that some of us pagan types give lip service to, paying homage to the long-gone hermit of Concord who said, "In wildness is the preservation of the world"? (And what exactly does *that* mean, by the way?)

There are multiple definitions of *wild*, including "lacking discipline, restraint, or control, unruly," and "full of ungovernable intense emotion," but these, and other frat boy definitions, are not what interests me. "Reckless, risky"? That's closer, as is "random or spontaneous." Wildness as spontaneous, random, unexpected. As the opposite of planned, determined, controlled, domesticated.

Here we may be edging closer to Flaco's deeper appeal. I think that there is in each of us a desire to break out of the lives we find ourselves trapped in (even when those lives are safe and, for the most part, good). Those moments when we finally do break out can loosely be called wild. We suddenly realize things aren't the way we have been telling ourselves they are. We are shaken out of ourselves. Jostled. The rattling hamster wheel of our minds suddenly stills. One part of wildness is surprise, the unexpected. But it is something else, too. Something ineffable. Something we crave but have forgotten we crave. Something we have pushed aside, decided not to take seriously, dismissed to our own detriment.

Bird-watching is usually not regarded as a wild experience, despite the places that birding can lead us. But while generally seen as a tame way to pass the time, bird encounters can lift our lives. Watching is not always just watching. You can be captured by the idea of flight, of lifting off, of leaving yourself behind. It doesn't hurt that this leaving is often accompanied by an internal lift, something close to what we once called ecstasy. Maybe ecstasy is going too far, but there is at least a brief release, an internal flight in response to the external one. Birds help us travel to places beyond the walls of the brain. We escape what the biographer Walter Jackson Bate called "the prison cell of self."

"Strange to have come through the whole century and find that the most interesting thing is the birds," the Cape Cod nature writer John Hay said to me once while we stood on a beach watching northern gannets dive from high in the air, slicing into the bay like living lawn darts. "Or maybe it's just the human mind is more interesting when focusing on something other than itself."

In turning our eyes beyond ourselves we find not an answer but a mystery, and perhaps an eventual understanding that all we are is a part of that mystery.

NAN AND I TAKE our drinks to the living room where the windows look out at Central Park and the reservoir. Across the park shine the lights of the Upper West Side, where Flaco spent his last days. We sit across from each other and talk for a while about books before getting back to birds.

"That brief encounter transformed into a deeper affection," she says. "I got up at 4 a.m. that night and went to the library and sat on the couch. I watched the Flaco videos from earlier in the day and looked at the pictures. I realized that I missed him so much. It was strange, but I thought: I've fallen in love with an owl.

"I missed that surreal, wondrous, exciting, and funny time with him. I still do. It was that strong, like being lovesick, and it remained that way for quite a while.

"This all happened right before Thanksgiving. And at Thanksgiving I'd catch my kids looking at each other across the room and almost laughing. And I would ask, 'What is it? What is it?' And they would say, 'Well, you sure talk a lot about Flaco.'"

I HAVE NEVER HAD an owl visit me at home, but I have had my own owl encounters.

I relish them.

A bird encounter is one thing. An owl encounter is another. I have studied raptors, ospreys in particular, for close to twenty-five years, and I thrill whenever I see one. But among bird encounters, owl encounters are unique. Stealth is part of that difference, and with stealth comes surprise. I remember walking down a sloping hill through the woods near the Wachusett Reservoir in central Massachusetts. The sun had just set, and darkness was coming in gulps, the fading light lending an eerie feel even before something, some presence, flew over my shoulder. Huge and silent, that presence clarified itself into a great horned owl. Its approach had been nearly soundless, and it had gotten very close, as if considering having me for dinner. After it flew a few feet over my head, it dropped further with the slope of the land, flying low as the trail led downhill.

Or how about this: One January day I was walking the beach near my house on Cape Cod, a beach so tiny and domestic it was locally known as "the little beach." It wasn't *little* that day. The bay was frozen for a quarter mile out, and geysers of salt water were spraying upward through cracks in the ice like the spray of spouting whales. As I reached the beach, I saw sitting there in the tundra, unperturbed, a messenger of the primal. A snowy owl, far from home, stared up at me with black slits for eyes, the ocean wind blowing its white feathers back like a boa. This was before the irruption of 2013, when snowys became a common sight on Cape Cod. The owl was less than ten yards away and didn't feel like moving, so it just sat there, looking both spectacular and perfectly at home on a beach that on that frigid day truly resembled the tundra where it spent its summers. Having come down for a visit from the Arctic it seemed to have brought that region along with it.

The owl appeared to be wearing a shining white robe, though the robe was flecked dark with brownish markings the color of chips on a cinnamon scone, likely meaning it was a female. As she swiveled, her great disc of a face transformed itself, one moment a white lion, the dashes of black for bill and eyes a mere cartoon outline, the next more hawklike and predatory as she lifted a hand-like talon to scratch herself, and then, when her eyes turned entirely away, a featureless white dome. Mostly she squinted in the wind, but when her eyes reopened, they shone yellow.

Her tail feathers blowing behind her like streamers in the heavy wind, the snowy seemed made for spectacle. She had *presence*.

SOMETIMES IT SEEMED LIKE the only person who really knew how Nan felt was David Barrett.

"I reached out to David. I could share the whole experience with him. And he immediately saw that I was in love with this owl. And so there was this immediate identification with each other. Every day David would let me know what was going on because every night he would be out there. Pretty soon we were texting all the time."

She began to follow his Manhattan Bird Alert and spend time on X following Flaco's followers.

"The first time I posted, I put up a picture of Flaco at my window, and the phone started going beep, beep, beep, beep as people responded. It was completely unnerving to me. Thousands and thousands of people were responding. I couldn't keep my phone anywhere near me. And so I deactivated my account with Twitter because it was making me nuts."

But she kept in touch with David, and he kept giving her daily updates.

"It was always me and David Barrett. It's so funny because David and I both stay up late. We discovered early on that we were both that way, that we were, you know, like owls. Yes, exactly, like owls. I usually stay up until two. And he apparently does, too. David and I, you know, there was this really intense relationship because of the sharing of the owl. And we never met until after Flaco died. We just texted. But I was feeling like he was a really close friend."

She puts down her drink and looks over at me.

I feel she wants to say more, so I nudge.

"He seems like a very careful person," I say. I choose that word, well, carefully.

"Yes, very careful. But also nonjudgmental. He was always nonjudgmental with me whenever I thought I had gone over the line during the media frenzy. At one point *Inside Edition* pulled back from a second interview with me. I wrote to him in a paranoid mood and said, 'I think people are really annoyed at me with all this playwright-poet stuff.' I was feeling like, you know, all these serious birders had been following Flaco for months, and now every time there was an article or TV interview here was this woman getting all these interviews, and that it might be really annoying. I was getting all this publicity, and I think that for some people it was like, 'Well, excuse me, I've had a relationship with Flaco for a long time and why is she getting all this?' And he wrote me and said, 'No, people love you and everything about the Flaco story.'"

Her mood changes from apologetic to mildly defiant.

"Well, you know why they all wanted to talk to me? Because the bird came to my house. I'm sorry it didn't come to your house, it came to mine."

Before I leave, we take one last look at the kitchen.

"I can't go by the kitchen window now without automatically looking, even though I know he's not there. It's this instinctive thing. For a long time I hoped he would come back. David used to say to me he might come back and so for the longest time I kept looking and thinking 'Please come back' and I still can't stop doing that, can't stop looking at the window. I can't look out of either window without sort of expecting and hoping. It's a sort of a denial thing, too, a disbelief in a way that I won't sometime see him out the window again. It's totally irrational, but I can't get over the visceral excitement and impression of looking out a window and seeing him."

I WISH I HAD seen him.

Our two species share forward-facing eyes. Owls are the only birds with binocular vision. Vision is vital for both species. We want to *see*. When Nan Knighton had stared out her window, Flaco had stared back.

What was it like to be in your kitchen, that most domestic of rooms, and be visited by the wild? A kind of everyday transcendence, a quotidian epiphany. The word *visitation* is defined, according to my dictionary, as "the appearance of the divine or supernatural." Flaco was not divine or supernatural, but for Nan this was surely a visitation.

Later she wrote this about the owl:

He'd been winging round the city all that time,
Mostly on the Upper West Side (though there was a rather mysterious trip to the East Village).
And then he came to me.

"He came to me." Silly. He came to a window ledge. *Any* window ledge. No. Mine.

He came to me. Hard not to be jealous. How we all wish he had come to us. But that is not the nature of these things. Not the nature of grace. Thousands went in search of Flaco, but Flaco had come to her. *Blessed* is not one of my favorite words, but it is a word that many people used to describe Nan's good fortune, and it seems to fit well enough.

5

A WALK IN THE PARK

THOUGH I AM not yet keeping owl hours, I think it's fair to say I am becoming crepuscular.

Last night at dusk I walked through the North Woods. All day I'd gone in search of things, part of a birding group marching through the park, but as the sun set in the evening things started to come to me. Seeing dozens of birds with yesterday's group couldn't match my experience at the end of the day when, alone near the Loch, a single black-and-white warbler with gleaming racing stripes flew right by my face. It had come to me! Such beauty in those oh-so-distinct and vivid stripes.

Not long after, I met up with a Flaco follower named Sandra Beltrao, who I had texted with earlier that day, and when she arrived, she told me about an evening the year before when, at dusk, she experienced Flaco flying right by her head in those same woods. She felt the wind of his wings as he passed.

Sandra loved that solitary moment, but she had also found joy in being part of the community of bird photographers in the park. "I sometimes liken us to a SWAT team that has to use the utmost stealth and accuracy to find a target bird," she told me. "And when we do clap eyes on, say, a beautiful Blackburnian warbler, there's a collective gasp followed by silence. At this point a lesser bird photographer will crack under the pressure, step on a twig and flush the bird, but you would have got your killer shot. Or else you hang back while the others move on, and you wait and wait and wait. Eventually the bird forgets you're there and turns into a little Caruso or Nijinsky."

This morning I begin my journey at dawn. Another month has passed since Flaco's death and spring has bloomed, and I start my walk by paying homage to the tulip tree where Flaco spent his first free night, across from the Plaza and Bergdorf Goodman. The display has been changed again at the department store; the headless mannequins on the left now seem to be inside a giant boom box, while the ones on the right with tufts of owl-like hair growing from their heads pose around a movie klieg light. I follow the usual Flaco route by the Plaza with its beckoning red stairs. Pigeons take off in front of me—only pigeons, but that sound of their wings beating all at once is something, and their cooing seems to throb up from the sidewalk, the whole sidewalk, maybe the whole city, cooing along with these supposedly unspectacular birds. The park is greening, far behind North Carolina where I watched the sun rise over Cape Fear just four days ago, but still beautiful, trees spraying and splattering a Seurat green, leaves that were bare when Flaco flew this route now on the edge of full bursting.

As I cut into the park and up toward the Hallett Sanctuary I pass a homeless man on a bench. Preoccupied with the business of keeping alive, he ignores me and stuffs a blanket into his pack, which holds his earthly belongings, but then he looks over at me, and at just that second a rat scurries across the path. We nod to each other in acknowledgment of the rodent. I think, *Flaco food.*

At least the man wasn't too cold last night. A truly perfect spring day, one that, as Henry Miller once said of a perfect nap, puts velvet in your spine. I cross the bridge toward Hallett, Flaco's first brief home. The skaters have put their skates away over at the rink. Spanish bluebell that looks like purple bell to me adorns the edge of the sanctuary, and a northern flicker shoots by, flashing its white tail. It is early, so Hallett is locked, as it was those first days as the zoo employees tried to capture the escaped owl, but the little sky island is so much greener than when I visited in February, the leaves filling out. Apple trees burst white, cherries pink, and surrealistic green everywhere. Behind the hill of Hallett the buildings rise with straight-backed arrogance, jutting skyward, a testimony to the pomposity of straight lines in contrast to the gnarl of an apple tree's trunk.

That very first Flaco morning dozens of bird-watchers were already staring up at Flaco as he perched on his tree in Hallett, and if you are of a certain mindset it is easy to be critical of their voyeurism, of the funhouse mirror aspect of the Flaco story. But after weeks of talking to those who followed Flaco, I am convinced of their passion. Many cried when the owl died. Many use the word *love*.

You can argue about whether Flaco was wild or not, but many felt a visceral connection. And those many were part of a community. Nan wasn't the only one who experienced the jolt of the unexpected. Of course, since we are human beings, we quickly make the bird fit our own narratives, reshaping his story into ours. Raw experience is hard on humans. We seek to control it, if only by our stories. Flaco himself was perhaps the only one who experienced real wildness, however briefly, though fear, trepidation, and anxiety were all part of that wild mix. It wasn't ever as simple a story as the one the newspapers told.

After Hallett I head north. Having gotten to know this territory a little over my last few visits I now move maplessly through the park. We are always surprised by the ability of animals to navigate their worlds, but why? Like us, they remember where they have been, only more keenly, and this is particularly true if the animal can fly above the landscape, looking down on it like a folded-out map. This acuity should come as no surprise: their lives depend on knowing where they are.

Light is returning to the world as I make my way north through the park, following Flaco's route over those first weeks. My next stop is the literary walk, the promenade between noble late-blooming elms and the scene of some of Flaco's earliest explorations. I pass a statue of Shakespeare, who stares broodingly down at some maroon-and-white tulips. Here Flaco flew among statues of women suffragettes and an ecstatic Robert Burns, staring up rhapsodic with a quill poised in his hand but no paper below and never writing, and a perturbed Francis Bacon, troubled no doubt by some existential dilemma as he wrestled toward his philosophy of empiricism. Since the idea of story-making is so central to my own version of the Flaco narrative, it would have been nice if the owl had landed or perhaps shat on the head of one of these writers, but there is no evidence, photographic or otherwise, of such a lucky event. The sun finally starts to break over on the park's east side, near Nan's apartment. A qualifying race for the New York marathon is taking place this morning, giving the park both a festival feel and the divine gift of porta-potties, solving one of the main problems of a daylong hike in Central Park. To get to the porta-potties, a line of seemingly hundreds of them, I pass by a statue of an angry Beethoven and then, fittingly, one of two eagles glorying over a goat they have killed (the oldest sculpture that isn't a monument in the park, I will later learn).

A kind of new dusk returns as I enter the shaded Ramble, its winding paths transformed by the gloaming, giving it the feel of a real woods. As Marie Winn pointed out in *Red-Tails in Love*, wilderness was not what Frederick Law Olmstead and Calvert Vaux had in mind when they designed the park. The Ramble, now the most closed-in and wild feeling of the park's geographic features, was originally conceived as a grassy open-aired area where the occasional peacock wandered by. Winn writes:

> The Ramble's transformation did not occur by plan. Neglect created the unknown meadows that soon became

sparrow heavens, the unpruned trees and dead snags offering prime real estate for woodpeckers and racoons—neglect dictated by New York's ever-increasing budget constraints.

As the metamorphosis of fake to real progressed, word began to spread among bird lovers: Central Park was a great birdwatching area.

The metamorphosis of fake to real. I like that.

It's a city of millions, but I'm alone in the woods. Soon enough, without trying very hard, I see another northern flicker and hear white-throated sparrows, common birds with a beautiful song. I pass below the exploding cherry trees, pink petals falling, and climb up some rocks not far from where a fellow human, one without a home, is camped out.

Other than the single sleeping man, I've been alone for most of my winding walk through the Ramble, but now someone is coming toward me. I notice that he has the same type of binoculars that I do, minus the duct tape, a bird-watcher obviously, and so I say hello and ask him the logical questions, the burning questions: Does he know of Flaco? Did he ever see him?

He says that yes, he does; yes, he has.

In fact, he was there the very first night.

It is one of those coincidences that always seem to happen when you throw yourself into stories. That is, when you get obsessed.

He tells me his name is Edmund Berry, and I laugh and tell him I recently quoted him in an article I wrote. "I read that article," he says. We laugh at the absurdity of meeting in the middle of the woods, and I ask for details about that first night.

He tells me he saw an X alert about a strange sidewalk owl posted by a filmmaker named Penny Lane, and he jumped right on the subway and arrived on the scene at around ten o'clock, maybe between ten and eleven, he corrects himself. He took one of the first pictures of the bird. He says it was confusing at first to see a large bird on the sidewalk, and he thought maybe it was a great horned owl and that it had been hurt and couldn't fly.

"I wasn't used to seeing a bird in that state. He was sitting on the sidewalk with the police and crowds around him. It totally takes you out of the situation. It's not like seeing a bird in a tree."

The police had put up their yellow barrier tape, and then they placed the cat carrier next to the bird, who just stared at it. When they put a larger carrier down—"one that could hold a golden retriever"—the owl flew off.

The rest of the onlookers dispersed after Flaco flew south, but Edmund thought, *I can't go.* He followed Flaco as the owl flew down to the Pulitzer Fountain and the Plaza and Bergdorf Goodman. Alone now, he stared up at the owl perched in the tulip tree. It was a cold and windy night, but he was transfixed. There was an intense period when it was just him and this strange, beautiful animal. Later David Lei showed up and joined him.

He watched the owl for an hour, then another hour. He thought, *Well, I'm just gonna stay here. I don't know what else I can do that's as awesome as this. This is so incredible.*

"I couldn't leave," Edmund tells me. "It took hold of me. It eventually took hold of everybody. I just had to watch him. And the whole city soon came to the same realization: We can't go. We have to stay. We have to watch and see what happens."

After we say goodbye, I find a small dark corner of the Ramble and sit on a rock ledge and stare at water running over stone. For a minute the creek has the feel of a creek out west, the light lending the place a green pointillist beauty. The flowers of a dogwood lift toward the sky, its cups reaching upward, teacups for the rain. A red-bellied woodpecker lands on a nearby tree.

I think: how lucky for Edmund Berry; for Nan; for Sandra, who had been close enough to feel the wind from Flaco's wings; for David Barrett and David Lei; for hundreds of others, thousands really, millions maybe, on the ground and online to have had Flaco come into their lives.

To be lifted out of the ordinary. Who wouldn't want that? Who doesn't hunger for encounters beyond ourselves?

We all want to touch a world beyond our own. For a brief moment.

It seems one of the best things on Earth.

To reach out of our caged world and touch a wilder one.

I EMERGE FROM THE Ramble and climb to the Belvedere Castle overlook, where I can look down at Turtle Pond. On the rock ledge directly below me is the mother Canada goose I saw yesterday. She is still doing just what she was doing then: sitting oh-so-patiently on three eggs. Behind me is the black alder tree where I saw the kinglets on my last visit and beyond it the phragmites where the coyote prowled. I am a newcomer here, but already stories are attaching themselves to places and, in one human mind at least, changing them. Following Flaco's year is helping me learn a new landscape.

I decide to dip back into the Ramble, perhaps hoping for another mysterious encounter. What I find instead is a large group of people armed with binoculars that they are pointing up into a slightly freckled London plane tree, a tree that looks like it is wearing camo. The man who stands in front of the group and directs their attention at a small bird up in the branches is Bob DeCandido, aka Birding Bob.

I know this because Bob was the leader of the tour I took yesterday, a tour just like the one I have stumbled upon. Yesterday morning Bob led me and a group of about thirty other folks, including David Barrett, through the same paths I've been walking this morning. In fact the difference between the walk I have taken so far today, solitary and contemplative, and the walk I took yesterday when I was part of a large group, mirrors a rift in the world of Flaco fans and in the birding community in general. Very roughly speaking, it is a rift between the populist and elitist takes on birding, though *elitist* might not be exactly the right word for the ornithologists and bird-watchers who prefer a quieter and less intrusive kind of interaction with birds. Maybe *purist* is better. If one were to caricature the two sides, it might help to turn to that cultural touchstone known as *Caddyshack*. The elitist birders would be embodied in Judge Elihu Smails, the country club snob played by Ted Knight, and they would look down on the populist birders the way the judge looks down on Al Czervik, the crass new money character played by Rodney Dangerfield.

The reality is a lot more complicated. For instance, Birding Bob, the classic regular guy Al Czervik birder, has a PhD, worked as a biologist for the NYC Department of Parks, and spent many years traveling internationally studying birds. But the criticisms of Bob from old-school birders are many. That he plays distress calls to draw in birds for his groups to see, that he has been known to shine a light on birds, especially owls, at night, and that he draws large crowds of humans to birds who might prefer to be alone.

This last criticism was also leveled at David Barrett during Flaco's days in the park. Were the constant postings of Flaco's location drawing crowds that put undue stress on the poor owl? And did those crowds actually impede, intentionally or not, the recapture of Flaco? Certainly there are those who think so and said so (and still say so) on X and elsewhere.

David would contend that everyone had a right to see the amazing sight of Flaco in the wild, and that those who saw him would become more attached to both the bird and the natural world. David might even see himself as an educator and a guide through a portal into a wondrous world. That there was a whole lot of ego packed into the equation, on both sides of the conflict, is a given.

The birding world has changed a lot in the last decade, and not just because of the internet and Twitter/X. COVID sent thousands of people with nowhere else to go outside. Some of those people found disc golf, others found birding. Central Park birding boomed, and David Barrett both helped create that boom and rode it. He broke the old rules.

In particular he broke the rule about revealing the exact locations of rare birds in general and owls in particular. This enraged those in the purist camp and led to large crowds where the select few had once tread quietly.

But David changed Central Park birding in other ways. He was a meticulous researcher and studied trends and maps and weather reports to anticipate where birds would be. He learned everything he could about the habits of the birds he studied, trying to think like a bird.

"He brought something new," another park birder told me. "Like *Moneyball* for birds."

During our chance encounter, I had asked Edmund about David. Edmund said he liked David quite a lot, and also thought he understood his method.

"He is a bond trader, or was a bond trader," he said. "I am now a bond trader, and I think I know the type. I saw what he was doing. I got it. He applied his knowledge and techniques to birds."

I asked Edmund about Bob.

"I'm a huge fan. Bob is a controversial figure in the park, but I think he is terrific."

I nodded. While I like quiet and walking alone in nature, I also like Birding Bob.

Yesterday, wearing a yellow baseball cap and wielding a voice that carries well, he guided us through what he called a subpar morning but one that for me was a relative bonanza: in short order I got to see a red-bellied woodpecker, a pine warbler shining bright yellow, a flicker in a cherry tree letting out its war cry, a yellow-bellied sapsucker, a brown thrush, and even a racoon climbing along some rocks. He pointed out a cardinal "for the Europeans," since what is a common bird for us is a rarity for them.

I arrived late, so I wasn't sure how Bob got paid for his troubles, though I saw a few folks handing cash to Deborah Allen, his partner in life and, apparently, business. Many of the birders were beginners who he'd lent binoculars to, and there was no self-consciousness among us when we asked questions that real birders might think inane. In fact, an early high point came when a few of us abandoned the search for ruby kinglets in the alder tree and pointed our binoculars at something more interesting going on down by the shore of the reservoir. A young man was getting down on one knee in front of a young woman, and then she was putting her hands up to her face. Soon more than a dozen of us were watching the action through our binocs, and after the couple hugged and kissed, a musician, who had been playing guitar during the proposal, yelled up to our group, "She said yes!"

When the other birders left, I, following the dictates of my nosy profession, wandered down and talked to the couple. The rock path that led to the water was strewn with rose petals. They told me their names were Andy and Sorcha, and no, they didn't mind if I wrote about them in a book. "Sorcha's an Irish name," the newly engaged young man said in an accent from the same country.

While David Barrett was part of the Birding Bob group, and indeed had been the one who invited me along, we didn't interact much other than his pointing out a bird for me once in a while. He was all business, and his business was taking pictures that he could post. At one point he showed me a photo of a duckling on his phone.

"I just posted this a few minutes ago," he told me. "I knew it would do well."

By "do well" he meant of course that it would get a lot of likes.

Once the tour wound down the few of us who were left headed over to the Reservoir.

Bob, who I could easily imagine bellying up to the bar with, and David, precise and sometimes remote, are a kind of unlikely pair, though they both fall under the broad "populist" label of birders. Bob is a staunch David fan.

"Nobody works harder than David," he said to me as we stood looking through our binoculars at a loon out in the middle of the reservoir.

At that very moment David walked up and Bob said, "You should have hired me for your publicity department."

Then, with David still next to us at the reservoir railing, he continued, "I'll say it again. You heard it. Nobody works harder than David does. Or is more competitive. This guy is a terrorist, man. You see him, neat and thin, sunglasses, and you don't realize that."

Soon there were only five of us left in the little group: David, Bob, me, a young birder from Thailand named Cyrus, and Deborah, who I had heard was an amazing birder in her own right. As we looked out across the reservoir, Bob managed to both keep a steady stream of words flowing in my direction and interject bird IDs—"There's a lesser scaup and some female buffleheads."

"People attacked David, the zoo attacked David," he said. "But the thing is, the thing that doesn't get told, is that David, through Manhattan Bird Alert and Twitter, developed a whole community of people who weren't birders before. And they would report, 'Oh, I heard Flaco on 103rd Street and Riverside Drive near my house.' So it became a thing to do. People got so into it.

"David takes a lot of flack from the serious birders, and I do, too. Oh, we're harming the owl. Well, we weren't harming the owl. As for the zoo, get rid of your public relations department. Get somebody new because you guys blew it with this. They want people to see and learn about animals? Well, this was their great chance. They could have been part of getting people out into nature and educating. Meanwhile David did the real work. He knew immediately when that bird showed up that it was going to be a big thing."

Bob paused to point out a red-breasted merganser before continuing.

"It's New York, so people live to fight. It's better than coffee. There's a schism in New York City birding. People like me and David, who work with people, break the rules. You know what I mean? If you go online, you can read the Audubon ethics. Don't use sound recordings in heavily birded areas. Don't do this. Don't do that. And I've just pushed on through it. Because what you do is you develop your own community of people.

"Early on I would do owl walks at night, and I would bring an owl along to attract other owls, and we would publicize their location. But then a new group formed, because they didn't want to have anything to do with David or me. And they weren't doing anything to make it more popular for people. David caught a lot of flack. That other group has always been looking down on the Manhattan Bird Alert and David. But David single-handedly made Flaco into a story. And the zoo should be ashamed. They should hide, the way they blew that story."

David had been listening with me, but now he interjected, "It wasn't good PR for them. They even tried to attack me. Another mistake."

His measured voice played in counterpoint to Bob's booming.

"A small minority always wants to oppose it," David continued. "It's true of anything, like with a popular singer. Much of the world loves, but a minority hates. It's just how things work out."

A popular singer. I found myself wondering if he could name one.

"There's a strong sense of righteousness," Bob added.

"Have you gotten used to it?" I asked David. "Have you developed a thick skin?"

"Certainly. It's the nature of social media fame. You take anyone who is a creator and gets a strong following. There will be other creators attacking that person all the time. It's quite competitive."

He added, "This one particular troll was quite intense about it. Obsessed in fact."

"I think I know that troll," I said. A few weeks before, I had studied all of the Flaco tweets over the previous year, and hers stood out for their viciousness.

"David is really good at being anonymous," said Bob. "He can back out of stuff. And he's really good at being nice. I'm not so good at that."

Deborah had spotted the loon out on the water again and said, "It's almost in full breeding plumage."

Bob then demonstrated a technique that the purists must surely look down their noses at. He played a loud recorded loon call, the haunting and well-known call emanating from his phone. Sure enough the loon started swimming closer. Bob kept talking while staring at the bird through his binoculars.

"The zoo got on Twitter and said, 'Please, we don't want this publicized.' Well you can't hide an owl in Manhattan. And then if people show up and watch him that's David's fault somehow? They said the people will impede capturing the owl."

"Did people actually impede the capture?" I asked.

"Well, they are standing around, a crowd," Deborah answered. "Maybe that makes it harder to get closer to the owl to trap it."

After a while we headed back to the Loeb Boathouse, where the tour started. Most of the way back Bob talked to Cyrus about all the times he and Deborah had birded in Thailand. "If you think New York birders are competitive you should see the Thai," Bob said to me. It turns out Cyrus and Bob knew a few people in common. When we reached the Loeb Boathouse, I said goodbye, and as he shook my hand Bob said, "I would like to see people go from Flaco to, 'Okay, what other owls do we have in New York City, and how can I see them?' I want to see that jump made."

THIS MORNING, BEFORE I started my dawn walk, David Barrett texted to tell me that yesterday an osprey had been seen diving for fish in the Harlem Meer, the man-made lake in the northeast corner of the park. *Meer* is a Dutch word for "small sea," and the name was also a nod to the formerly Dutch village of Harlem. I have never been to the very northern edge of the park before, and I make my meandering way there, circling the reservoir beneath the cherry trees while watching buffleheads surface dive into the clear water. Clear water in the middle of New York? Yes. In terms of visuals at least. I am able to follow the buffleheads fifteen feet below the surface as they tunnel downward. In a way it is like swimming with them.

After the reservoir, I take a detour back to the North Woods to pay homage to two of the roosting trees, the famous black oak and the white ash. I stop in at the compost pile where Flaco spent so much time, the dug-up earth being good for rats, and stop again to talk to a couple who are throwing a Frisbee for their Weimaraner, finally asking if I can throw one for him before moving on. (It is a well-preserved ultimate disc from 1979, the year I started playing the sport.) It's Sunday in the park and a perfect spring day, the apple, pear, and cherry trees bursting along the Great Lawn, and the dogwood everywhere adding a startling white against all the new green. I follow a road that runs north and downward, alongside the massive construction site—they are rebuilding the large public pool/rink and the area around it—where Flaco liked to perch atop the cranes and tractors.

There are plenty of birds at the Meer—grackles, cormorants, red-winged blackbirds, and swallows carving up the air—but other than the swallows, the skies up high where raptors would soar are empty. Standing below a bald cypress, its knees rising from the brown water like sand doodles, I don't see much chance of seeing an osprey before I have to rush back to the hotel and then rush to my plane. You can't force these things, I tell myself. I circle the Meer but then plop down on a bench, resigned to the fact that there will be no ospreys today. Isn't wanting more a little greedy of me anyway? The fact that I ran into Edmund in the Ramble has already elevated the day above an ordinary one.

I know I'm a bumbler as a birder. Too impatient, for one thing. But I also know that as an osprey watcher I'm not bad. It all comes down, as so much does, to obsession. You can get good at things by dipping in. But great? In my experience the only way to approach that is to make it your everything, at least for a while. For one year of my life I made ospreys everything. I started from nothing. When I began watching ospreys I couldn't tell them apart from a gull. After a year I would joke that if a giant osprey were discovered hurtling through space toward our planet and the president called in Bruce Willis to gather a crack team of a half dozen osprey experts from around the world, I would be chosen as the lay expert. When I started, I hadn't witnessed a single osprey dive, but over the next few years I saw hundreds of dives in dozens of places, up and down the Atlantic coast, in the Rockies and West Coast, as well as in the waters off Cuba and on Venezuelan rivers. "When we try to pick out anything by itself, we find it hitched to everything else in the universe," wrote John Muir. I got lucky and picked out ospreys.

Today my luck continues. I am still sitting on my bench, resigned and ready to leave, when a boy and his dad walk by. Their field markings—binoculars, a telephoto lens on an expensive camera, questing looks—reveal them as my fellow bird people. So I start up a conversation with the dad, explaining that I am here to look for ospreys, and, lo and behold, they are here for the same reason, having seen yesterday's posts on X about diving ospreys. Though the last thing my story needs is another David, the facts can't be ignored, and it turns out that's the father's name. David *Bellass* is from the UK, but he now lives in the States. His son's name is Alfred; he is thirteen and charmingly shy and polite. Since they are going north around the Meer and I am about to head south, we exchange phone numbers so we can text if we see an osprey. I sit for a while more, watching a bossy Canada goose practically land on top of a row of cormorants, then push off. My mind is taking the Uber to LaGuardia, standing in the endless security line, making my ragged way home. It is obvious that my glorious day in the park is over.

I'm already living in the future when my phone dings. I think it might be the new David, but it is an old one: David Barrett.

His text message holds exciting news: "A birder told us there is an osprey perching on the island on the west side of the Meer."

Before I start hustling back, I copy the text and send it to the new David.

By the time I get back, David and Alfred are staring up at a large bird with a white head and a black mask perched in a tall tree. I high-five with Alfred. Another birder named Stephen is there too. We watch for a bit, happy to have achieved one of the day's goals. After a while the osprey flies off toward the road I descended earlier, the one by the pool construction site. I say goodbye and follow the bird on foot alone while Alfred and his dad keep watch on the island. My search is futile, and soon I have circled around to the other side of the Meer, almost directly across from David, Alfred, and Stephen, who are peering up at the sky through their binoculars. They text me that the first osprey has returned and there is now a second osprey, and they have seen it dive. A moment later the bird flies directly over my head, fish in its talons. I expect it to land in a tree and tear right into its fresh meal, but instead it flies out of the park and over a building, banking south, seemingly flying deeper into the city. Without thinking I text David Barrett about this peculiar behavior.

It is really not so peculiar, which I would have realized if I had stopped to think for a minute.

"No these birds are nesting now!" he writes me back. "The fish must be for its mate. Who needs to stay on eggs."

"Of course!" I text back. It's embarrassing. I am out of osprey shape, and I have committed an osprey faux pas, or at least revealed that I am not the expert I claim to be.

I consider my situation. At this point it will be challenging to make my plane, and anyway I don't really want to leave. Despite my technical ineptitude, I manage to use my phone to change my flight to the next afternoon, buying myself another morning of birding as well as a possible brunch with Hadley in the East Village. Stephen wanders over to my side of the Meer, and I bird with him for a while: he has Rasta dreads and what I perceive as a calm and maybe slightly more inward-turned approach to birding. I ask about Flaco, and he tells me he preferred to go it alone and that while he loved Flaco, he didn't like to approach him when the large groups were out. I appreciate that.

After a while we walk around to the other side of the Meer and reunite with David and Alfred. They are thrilled with how the morning has gone. Alfred is shy, but at his father's urging he shows me the photo he took of the second osprey diving, one of the best and most dramatic osprey photos I've ever seen. The picture, all abstract angles perfectly framed, shows the bird right before its impact with the water, its head down and legs thrown out to snare the fish. Later I will post it on X, and David Barrett will repost it, giving Alfred a little fame among Manhattan Bird Alert's 91,700 followers.

The next day I will text David Bellass and tell him what a pleasure it was to spend time with him and Alfred. He will write back: "Yes, a magical morning. My son was beside himself with joy and I was too."

Populists, purists. Those who prefer to bird alone, those who love groups. Beginners, experts. There may be rifts, and New Yorkers might like their fights as much as their coffee, but for me today has been a lesson in the way this community, whatever else it is, is connected. And now I am part of it, too. It feels like a web in the best sense. My text *from* David B., which led to my text *to* the other David B. and his son, was what led to our magical morning.

Now that I no longer have to catch a plane I decide to extend the magic. Following the web of connections again, I head north into Harlem toward the apartment of Ruben Giron, who Sandra put me in touch with, another example of the Central Park web. When we texted earlier, Ruben invited me over. Just like that, I'm a part of the group. Following his directions I walk for twenty minutes to his place, and when I get there, he buzzes me in and I climb up to his second-floor apartment. Soon we are sitting at his kitchen table and he is drinking a cup of tea while telling me about his relationship with Flaco.

"I'm not a birder but I love going out into nature. I'd heard about Flaco but hadn't seen him, and then one day he flew up to what's called the High Meadows in Central Park, and I just said, 'Oh, I've got to go see him.' So I was probably thirty feet away from Flaco. He was on a tree and it was a nice, bright sunny day, and I got my first shot of Flaco."

He describes the photo and then imitates Flaco, closing his eyes and tilting his head back to the right, smiling slightly.

"He was obviously basking in the sun. And I just felt this burst of emotion. It was like, wow, there he is, just enjoying the sun, letting the sun hit him in the most natural environment and just being there peacefully."

"I had heard about crowding of the owl, and it worried me," he says. "But what I found that day was that everybody had the best etiquette of being very, very quiet. Nobody was making a peep. And I felt like I was in the presence of this magnificent being and watching him enjoy freedom. So for me, Flaco quickly became a huge symbol of freedom, kind of beating the odds, even though with his death, some people are just saying, 'Well, that was expected,' and all of that.

"I don't look at it that way. I have a different perspective. Every living being is going to die. I believe Flaco enjoyed his freedom, and he lived it in the way that an owl will enjoy living. He used his intelligence to find places to hide, places to hunt and he did that, and he's a hero to me. I mean, it's what we're all trying to do. We're all living our life, trying to find that place of joy in life. Some of us are in cages. Absolutely the biggest cage is our own mind, but certainly being under the precepts of society and culture, there is an entrapment that occurs. We believe that we're supposed to be a particular way, but Flaco, he could only be Flaco, and he was enjoying that freedom."

I'm not surprised when Ruben says, "I'm a spiritual person." With his bald head and calm, smiling manner there is something monk-like about Ruben, and when I say this out loud, he confirms that this impression is not inaccurate.

"Birding for me is a way of meditating," he says. "I spent sixteen years in a yoga ashram, studying how to find my own inner joy. And when I see birds in the park, I feel joy, but the fact that I am using an external source to bring out that joy is really wonderful."

Focusing outward, not inward, on an "external source," gets at something I have been trying to articulate.

I ask about Flaco's flight to the Lower East End and his return to the park.

"It was an anxious time. He was gone for a week, and we thought he might have died. Then someone contacted David Barrett and said he's down in the village. It was a tremendous relief for me to know that he was okay.

"And that's when my feelings changed. Until then I was feeling that Flaco was going to be in the park all the time. And I believed, even though he would choose different locations, that I could see him again tomorrow or the next day or the next day. But when he left it added a new feeling of impermanence. And that impermanence really impacted me, teaching me to really value his presence in a more special way. And so he came back and everybody was like, he's back, but it was only for a day or two. He was there one day and then gone, and I happened to get pictures of him on that one day that he came back. I didn't look for his location online that morning. When I go birding, I go by gut instinct. He had been gone awhile, and I had no reason to think that he was there. I was down by the Loch, across from the compost, and I just said, 'Let me just go check and see,' and there he was. And I got some beautiful pictures of him in his tree, and I said, 'Oh, he's back.'

"But the next day he was gone again. And after that we saw him only periodically because he headed over to the West Side, you know, and he was hanging around the buildings and the water towers and fire escapes.

"And I thought that was really because he found a place where the red-tailed hawks wouldn't bother him. And away from crows. The blue jays, too, and even some of the smaller birds. You would see some of the smaller birds come and start pecking at him, you know, trying to chase him away. He obviously had a bird's intelligence, an owl's intelligence."

Ruben dips into another room, and when he comes back, he shows me a series of beautiful photos of Flaco, including the one of the owl basking in the sun. Before I leave, I thank him and ask him about Flaco's death.

"When I heard that Flacco had died, I had tears. And I felt like a very good friend of mine had just passed away. Yeah. I felt a lot of sadness, and at the same time, you know, it was just his time."

Flaco in his Central Park Zoo enclosure.
PHOTO BY ELENI PALMOS.

Flaco's first hours free on Fifth Avenue.
PHOTO BY EDMUND BERRY.

Flaco and a cat carrier on his first night.
PHOTO BY EDMUND BERRY.

Flaco in February, weeks after his escape from the zoo. PHOTO BY DAVID BARRETT.

Flaco perched on equipment. PHOTO BY DAVID BARRETT.

Early days of freedom. PHOTOS BY RUBEN GIRON.

Flacomania rising. PHOTO BY FERNANDA VASCONCELOS.

Close call with the snare trap. PHOTO BY DAVID BARRETT.

Flaco in May 2023. PHOTO BY DAVID BARRETT.

One of Flaco's first meals outside of the zoo. PHOTO BY RUBEN GIRON.

Basking in the sun. PHOTO BY RUBEN GIRON.

Flaco in flight. PHOTO BY ANKE FROHLICH, LINKTR.EE/ANKEFROHLICH.

Flaco snoozing. PHOTO BY SANDRA BELTRÃO @SANDRABIRDLOVER.

Flaco on a summer night. PHOTO BY DAVID BARRETT.

Flaco with fans. PHOTO BY ANKE FROHLICH.

Flaco with a rat. PHOTO BY DAVID BARRETT.

Flaco left the park on Halloween and made the city his home. PHOTO BY ROBIN HERBST.

In the city. PHOTO BY DAVID BARRETT.

Right, top: Visitation. On November 14, 2023, Flaco showed up at the kitchen window of playwright Nan Knighton. PHOTO BY NAN KNIGHTON.

Right, bottom: A view from Nan Knighton's study. PHOTO BY NAN KNIGHTON.

Flaco hits the city. PHOTO BY DAVID BARRETT.

Flaco perched on a rock at night. PHOTO BY ANKE FROHLICH.

On the baseball backstop of Hecksher Ballfields in Central Park. PHOTO BY ANKE FROHLICH.

Flaco emotes. PHOTOS BY RUBEN GIRON.

Above: West Side Story. In his final days Flaco called the West Side his home. Flaco atop Art Deco building at 241 Central Park West. PHOTO BY DAVID BARRETT.

Water towers proved one of Flaco's favorite perches. From there he could look down on the city. PHOTO BY DAVID BARRETT.

Flaco atop a Upper West Side water tower.
PHOTO BY ANKE FROHLICH.

Osprey entourage in Central Park. Left to right: author David Gessner, Alfred Bellass, David Bellass, Stephen Rogers. PHOTO BY SANDRA BELTRÃO.

Osprey Dive. PHOTO BY ALFRED BELLASS.

The male eagle-owl on Eagle-Owl Hill in Finland.
PHOTO BY HEIKKI WILLAMO.

The young birds in Finland. PHOTO BY HEIKKI WILLAMO.

Flaco by moonlight. PHOTO BY ANKE FROHLICH.

6

THE VIRTUAL BIRD

DURING THE WINTER of 2013–14 I got some time off from teaching and drove up from North Carolina for a twenty-three-day stay on Cape Cod. I was there to finish a novel I had begun to write more than thirty years before when I was living on the Cape. "Taking a book off the brain," was how Melville described the last draft of *Moby-Dick*. This was a process that took Melville a period of months, not decades, but I was hoping for something similar, in technique if not result. In a sense I was back there to finish a job. A wild job, but a job nonetheless.

The writing marathon was expected, more or less, but what lifted the trip into the unexpected were two gifts from above. The first were the owls. As it turned out, I was not the only one who was visiting Cape Cod during those winter months: an irruption of snowy owls had occurred, which meant that one of the most beautiful, stunning, and startling of birds, mostly rare in the lower forty-eight, had descended on us in numbers that were at once thrilling and puzzling. It was one of the deepest and strangest periods of nature immersion in my life.

Snow was the second surprise. More snow than I remembered during all my winters on Cape Cod. I was house-sitting for Katy Sidwell, an artist—extravagant and generous—and Steve Sidwell, a former defensive coordinator for the Patriots, Seahawks, and Saints, and Buddy, the rescue dog they found in New Orleans after Katrina. While the Sidwells were away, I walked Buddy, watched their art-filled house, fed their cat, soaked in their hot tub. On my third morning I was bird-watching in the hot tub, staring up at three titmice when a Cooper's hawk came diving after them. As I was looking up something came floating down, a flake landing on my cheek while I steeped in the near-boiling water. That was the first flake of millions. Before leaving North Carolina I had thrown my cross-country skis in the back of the car at the last minute. Now, each day after I finished writing, Buddy and I would push off up the long driveway and down a back road to the woods through an abandoned summer camp. The snow was great, with a good base, and kept getting better.

Which made it perfect weather for owl-watching. The owl irruption was a historic one in the Northeast and Great Lakes region, the largest in a century, and owls were seen as far south as Florida and Bermuda. A popular owl watching spot on Cape Cod was West Dennis Beach, and one morning, as the flakes fell at the start of our second major storm, I watched as an owl alighted on an osprey nest built on top of one of the blue boxes for trapping greenheads. Standing atop the huge nest, only six feet off the ground, the female owl looked startling and unafraid. Amid the sere colors of the marsh her whiteness jumped out, though the next day, with the blizzard's help, her camouflage would work perfectly.

The very best moment of the trip came later in the week when I was standing alone out on Coastguard Beach—on almost the exact spot where the naturalist-writer Henry Beston lived for a year in his outermost cabin in the dunes—when a young snowy owl rose off the tundra-like marsh with a black duck in its talons. The duck hung down limp below the owl and below that the duck's lifeless feet hung down even lower like damaged landing gear. The owl flew over the dunes with its prey, wanting to be alone, and I felt something of the irritation it must have felt toward me when I saw a couple climb over from the beach and attempt to follow it. We both, owl and man, were no doubt experiencing some variant of the same basic thought: What are they doing on *my* beach, in my *territory*?

My irritation faded when I caught up to the couple. Bundled up like refugees against the cold, the man and woman were considerate, giving the owl plenty of space. They were not birders, didn't even have binoculars, but they were so delighted by what they had seen I had a hard time being grumpy. The woman's face was radiant. *Radiant* is a word I used a lot during my previous few days of owl-watching, though mostly to describe the white unworldly shine of the snowy owl's feathers.

"It's only the second time I've seen an owl in the wild," the woman said.

She acted as if she had witnessed a visitation, which she certainly had.

"They're amazing birds," the woman said.

I told her I couldn't agree more.

We kept a lookout for another flight up from the dunes but decided to give the owl a wide berth as it continued its repast of duck.

HUMANS AREN'T THE ONLY animals that focus on owls. Other birds watch owls, too.

Mobbing is the term used for a group of animals harassing a larger predatory animal. Birds, being the one wild animal most of us actually see on a regular basis, provide the easiest way to observe mobbing behavior. It is likely you have watched crows or other smaller birds harass a hawk. There is risk for the smaller birds in this since one-on-one they are no match for a raptor. But there is potential reward as well: keeping the predator away from their offspring or from a food source.

The mobbing cries of the smaller birds serve to both call in more troops and add to the harassment. They also announce to the world: here is the culprit! This announcement undermines one of the predator's great weapons: stealth.

In her essay "Owls," Mary Oliver writes of a dozen crows harassing a great horned owl as it tries to sleep in the day. They "gather around it and above it, and scream into its face, with open beaks and wagging tongues." Oliver continues:

The caught crow is a dead crow. But it is not in the nature of crows to hide or cower—it is in their nature to gather and to screech and to gamble in the very tree where death stares at them with molten eyes. What fun, to aggravate the old bomber! What joy, to swipe the tawny feathers even as the big bird puffs and hulks and hisses.

It isn't only bold birds such as crows but smaller ones such as titmice that mob. The calls and the mobbing make what was private public, and so in a way mobbing birds serve a purpose not unlike the media. They draw attention to the target bird, making it a local celebrity. It is fair to say that among raptors, unlike among humans, all raptor celebrities are reluctant ones. For very practical reasons, they prefer to live incognito.

You have no doubt made the jump before me. Birds weren't the only ones mobbing Flaco. At the zoo he was at best a secondary attraction, an afterthought for the kids exiting the penguin and seabird exhibit. But in the park he was a cynosure, a star, a sinkhole, the bull's-eye of media attention. Everywhere he went people followed. The question is, how much did human mobbing interfere with Flaco's existence? Did we harass him as much as or more than the crows?

IN MID-MAY I FIND myself in New York City again, this time to pack up my daughter's dorm room and drive her back to North Carolina. We stuff her belongings into large blue IKEA bags that fill the back of our car. This is our fourth time doing this coming or going since she started almost two years ago, and it has become a ritual: being a dad, packing the car, unpacking the car, waiting down by the curb and maybe talking to other waiting dads while the girls are upstairs packing up the room, and moving the car when the garbage truck with the Yankees logo inevitably bullies through. Across from us is the Smoke Zone convenience store and the Bubble Tea shop. We've spent so much time in New York that I'm pretty sure Hadley is getting sick of us. She is our only child, and I will admit that we have had some difficulty letting go, especially since she picked such a seemingly dangerous place to go to school. Earlier in the month there was a rash of strange attacks near her dorm in the East Village where men ran up to young women and punched them in the face. Hadley managed to avoid those attacks but had almost been the victim of another in Union Square Park, when an obviously unbalanced man chased after her for blocks.

Since both my wife and I are writers, we have another curious way of not letting go. It can't be just a coincidence that both of us, neither of whom have ever made New York our subject matter before, are setting our current books in this city. My wife's is a historical novel that takes place in New York in the early 1900s. As for me, it is becoming more and more apparent that my next book will be about an owl.

Hadley turned twenty-one this Tuesday, buying her first legal drinks at a bar on midnight of her birthday eve. We drove up from North Carolina on Thursday, and after packing the car we will drive all the way back on Saturday. Nina decides to take Hadley out for a late birthday dinner after packing up her room on Friday, but I opt out, telling Hadley I will take her out the following week back in North Carolina.

Instead of joining them for the celebration, I am pulled, like a steel ingot by a magnet, back to Central Park. I have come to attend another birding walk, though this one will be a quieter walk than the one I took with Birding Bob. The Linnaean Society, founded by amateur naturalists in 1878 and including John Burroughs among its original members, is an organization originally devoted to the study of natural history in general, but over the years it, and its members, have become increasingly focused on birds. My host, and one of the tour's guides, is Alan Drogin, who lives in the building above the courtyard where Flaco was found dead. In fact it was Alan, the resident birder, who the super went to find when he discovered a huge bird lying face down on the concrete.

Before the walk, I visited Alan's apartment. I was apprehensive about the visit based on our email exchange over the last week. Of David Barrett he wrote, "Mostly I know about him from past controversies which I've avoided getting involved in. I don't use Manhattan Bird Alert, and the less said the better."

Based on his email, I was expecting someone on the extreme Ted Knight end of the *Caddyshack* birding scale. But Alan, a retired computer programmer with a gray Vandyke, thick black glasses, and an overall look even more professorial than my own, couldn't have been more gracious. I sat in a stuffed chair across from the couch where he sat and sipped coffee, and despite his email warnings, he almost immediately started talking about David Barrett. His take was nuanced rather than sharply judgmental.

"Personally, I would rather find birds on my own rather than going running after birds other people have found," he said. "Bird twitching is like an opium for people who are into that drug of chasing things."

Twitching is the word, more prevalently used in England, for running around and adding species to your life list.

He told me he had friends who were critical of David Barrett but that he had decided, "I'm not part of any of this debate."

His one point of contention, and one that had been the center of the Central Park debate long before Flaco appeared on the scene, was the revealing of exact owl locations, which was a breach of old-school birding etiquette. The rule was you did not give up the locations of owls and goshawks.

"Goshawks you can understand because goshawks have incredibly high prices on their heads, prices put there by wealthy people like Arab princes. They have detectives for all I know who will track this down and actually try to capture goshawks.

"So it makes sense with goshawks, but owls are a little bit more debatable. Nocturnal owls need to sleep during the day. And if you bother them too much, they're not gonna get a good night's rest. But even that is arguable. It's not like Central Park is great habitat for owls. If owls are getting bothered, they'll go elsewhere. They've got plenty of other places to go. And Flaco did, he did go elsewhere. Eventually."

David Barrett had defended his philosophy of owl-watching in his book, *A Big Manhattan Year*: "As I see it, alerting all local birders to an owl's location would do little to bother the owl while bringing much joy and satisfaction to the many people who care about the birds. I accept that there are instances where an owl chooses a low, exposed roost that can only be viewed from nearby, and then it is right to ensure the owl's safety by not broadcasting its location. I certainly do not want to see owls, or any sensitive species, harmed by excessive human curiosity."

I mentioned to Alan that early on in the Flaco saga David Barrett had said he felt that it would be best for Flaco to be captured.

"That surprises me. If that is true, the irony is that giving the exact location of wherever Flaco was, he was helping all of those people who were preventing the owl from being caught, and kept showing up and making sure he didn't get caught. He was drawing people there that were getting in the way. Even though he believed he should have been recaptured, he was actually undermining the whole operation. Why? Because of his belief in telling the exact locations of where it was."

Now, having taken the subway here from Alan's, we meet the other birders on the walk in Strawberry Fields, the five-acre section of the park dedicated as a memorial to the lost Beatle (and New Yorker) John Lennon. I shake hands with the small group of men and women who have braved the rain. The Linnaean Society in some ways defines itself against the Birding Bobs of the world, and it's true that the basic style and feel of the walk is different than the one I took with Bob. For one thing it is a lot slower. Due to the rain, the numbers are low, and many of the members of our group are actually group leaders, that is, experts, while Bob's group had more than a few beginners. It takes us about twenty minutes to go a hundred feet, and as we do members of the group call out birds—"There's a chestnut at three o'clock in the big oak." "I hear a parula." "There's got to be an ovenbird in that brush."—while I keep quiet.

We are really no less a mob than Bob's group. But perhaps we are a gentle mob, and perhaps this is a kinder and gentler sort of birding. Still in many ways what we are doing is not so different than Bob's tour, with the most obvious exception being that no distress calls are made to lure in birds.

But even here the lines are blurred. I note some pishing going on. *Pishing* is when a birder makes a sibilant *pish-pish* noise that simulates a bird's distress call, which can draw birds close. Fittingly, it's a noise that has the same mushy aural feel as the words *shush* and *tsk-tsk*. This, and bird imitations whistled or sung, are accepted in most birding communities. Is this really so different than Bob playing a recording? Both essentially trick birds and alter their behavior. There's also this: while David Barret may not be a favorite of this group, some members will admit they have benefited from his alerts.

I am exhausted after the long drive from North Carolina but resist the urge to retreat to our hotel and instead stay with the group. We barely move ten feet before stopping for the next bird, though I have a hard time following where they are pointing. I mostly nod and pretend to see what the others are seeing. The sole exception is a beautiful Baltimore oriole, its chest blazing orange when the sun breaks through, up in the top of a massive oak tree.

Central Park may be in the middle of a city, but it is also in the middle of the Atlantic flyway. The list for the walk will end up being a not particularly impressive one for a group like this, but it does give you a sense of what you can see in the park, even on a so-called bad day. We, or rather they, see rock pigeons, mallards, mourning doves, chimney swifts, herring gulls, cormorants, an egret, a red-tailed hawk, a red-bellied woodpecker, a northern flicker, a blue-headed vireo, blue jays, ruby-crowned kinglets, blue-gray gnatcatchers, house wrens, starlings, catbirds, veeries, a hermit thrush and wood thrush, a house finch, white-throated sparrows, grackles, red-winged blackbirds, a northern waterthrush, a common yellowthroat, a northern parula, a magnolia warbler, a chestnut-sided warbler, an American redstart, a black-throated blue warbler, a Canada warbler, northern cardinals, rose-breasted grosbeaks, house finches, and a robin or two, as well as the beautiful Baltimore oriole and an ovenbird (which did turn out to be in the bush).

At one point in the walk I stop to call Hadley and Nina, assuming the group will continue inching along at the same glacial pace. I am wrong. When I hang up, they are gone. *Poof.* They must have suddenly accelerated, off to some other spot in the park, and though I try to track them down for half an hour or so, I can't find them. I am on my own and tired and, as it turns out, somewhat sick of watching birds. The rain picks up as I retreat to the south end of the park and descend into a now-familiar subway station.

IT WAS ONLY LATER that Alan would let me know that among the birders in our group that day was another important player in the Flaco saga.

D. Bruce Yolton was one of Flaco's most articulate and consistent followers, regularly posting his thoughts on the famous owl on his blog, *Urban Hawks*, where he followed Flaco and wrestled with the issue of how the bird's popularity impinged on its health and safety, not to mention its (and his) peace of mind. If you never saw Flaco and want to get a sense of what it was like to follow him through the year, you could do worse than working your way from February 2023 to February 2024 on Bruce's blog. It is an extremely thorough and beautiful journey through Flaco's year, consisting of words, photos, and films. You can watch the year circle around from the coldest days of winter to the leaf out of spring to lush summer and edging toward the cold of fall.

The closest I have come to my own Flaco year was back in 1999, when I got my first-ever book advance for my book about ospreys. The advance wasn't much, but it was enough to quit my crappy job and buy a telescope and spend the better part of my days trudging out across marshes, beaches, and parking lots to four different nests and, to the best of my amateur abilities, observing four different osprey couples as they went about the business of rebuilding their nests, copulating, sitting on their eggs, and bringing in fish to feed their little reptilian-looking young until they grew too big for the nest and finally fledged. Osprey expert Alan Poole, who would serve as my Obi-Wan Kenobi during that year, told me I needed to live "on osprey time," and I tried my best, spending hours out on the marsh staring up at the birds, battling my natural restlessness and impatience. I didn't feel I should start writing the book until the ospreys migrated south in the fall, which meant that for six months I had to hold back my natural tendency to just dive into the work. Instead I watched as the birds flew, dove, fed their young, scratched, fidgeted, fought, slept, and soared. It was boring. It was exhilarating. I made my whole life revolve around the birds, and it changed me. At one point I saw someone else with binoculars hanging around their neck walking on the path back from one of "my" nests and I looked at him with gleaming eyes that all but said, "Have you found ospreys, too?"

Of course I missed my chance to do that with Flaco. It was too late. But following Bruce's blog through the year and watching the films he frequently took has proved a good substitute. Each morning these days, after I get my coffee and stretch my back, and before I settle in for a few hours of banging away on the computer, I watch Bruce's films and read his blogs and in this way recover a little of the sense of what it felt to watch an owl the way I once watched ospreys. I sip my coffee and watch Flaco, his proud yellow chest, the flecks on his wings like black paint marks, his orange eyes. These are not exactly action films mind you, no car chases here, but if you can work past your initial impatience, you get to something close to what I got to while watching ospreys. There is Flaco in his first free days in Hallett, practically blown out of his tree by the cold winds, his ear tufts almost horizontal, the branch he sits on swaying in the wind, moving his head in an almost catlike way but then swiveling it as if it were separate from his body in a way no cat ever could amidst the city noises of ambulances and horns and carriage horses clomping and dogs barking and planes flying overhead and of course those cameras clicking away that render his sanctuary anything but serene. And there he is a couple days later, putting his whole body into his first heard hoots, his tail bobbing and white chest feathers expanding.

That you have to slow yourself down to watch these fairly long videos and ignore the demands of normal life is partly the point. You have to live on owl time. And of course what is taking up twenty minutes of your life took up many hours of Bruce's.

Though he was an almost daily watcher himself, from almost the very start Bruce expressed frustration with how some people were watching Flaco. Here he is after the very first day:

> [Flaco] got harassed by a Cooper's Hawk and two Red-tailed Hawk. But it also got harassed by humans. An individual with no regard for the welfare of the owl jumped a fence into the Hallett Nature Sanctuary, and proceeded through the landscape in clear violation of the regulations for the sanctuary, which require patrons to stay on the paths. His response to being thrown out, was that it was worth it because he got great pictures. He went up to the group of us photographing the owl from across the pond, and said, "Can you believe it, someone called the cops on me." It made my day to say, "It was me." The sanctuary was closed to give the owl a safe space to come lower, and

hopefully be captured. No picture is worth putting this owl at more risk!

This is a consistent theme of the blog, as well as a growing antipathy toward David Barrett and his tweeting of Flaco's location. Here is Bruce on February 15:

> Thursday was an interesting day to watch the owl as it became apparent by the end of it, that he had fully adjusted to being in the park. But it was also a frustrating day, due to inappropriate use of social media, which hampered rescue efforts. I birded as usual starting at the reservoir and in the late afternoon went down to the south end of the park. With good luck and some intuition, I found the owl fairly quickly. I then noticed two photographers under the roost tree, and a bit later two Wildlife Conservation Society employees discreetly watching over Flaco. Everything was quiet and relaxed. Flaco was preening now and again. He looked like he was in the right spot, in a conifer near exposed rocks. This is similar to his mountain habit in Eurasia. Then out of nowhere, 15 people arrived. None of these people were the "regulars," who watch owls regularly but people I had never seen before. They quickly ran under the tree and all around it. Soon more arrived. It was a social media driven flash mob. I checked my phone, and guess what there was reporting of the roost on the Manhattan Bird Alert! After fly out, Flaco took a huge tour of the park's lower section stopping in many trees. Having watched owls be chased by crowds before, he clearly was being pushed by the crowd, who

were rushing to be under him at each tree. Eventually, as the crowd dispersed, he settled down.

And here he is on August 23:

I arrived after Flaco's fly out. He had a bit of an adventure. Some young men, who just happened to be walking by before he flew out ended up chasing after him to get cell phone pictures. A few of the regulars spoke to them, but Flaco wisely hid for about half an hour.

Flaco has attracted a number of inexperienced birders. When watching owls or any bird for that matter, you want to be careful about how you follow them. There is an art to following a bird. When birds move to a new location, they need to settle down and do an assessment of their new location. After a brief amount of time, you look at their behavior. If you see them alert and nervous, you keep your distance. If they look relaxed, you can slowly approach them. However, if you rush after them, especially if they are perched in a low branch or have gone to the ground, they will just keep going farther away.

So, if you're a new or inexperienced birder and visiting Flaco, don't rush after him to get a cell phone picture. You'll only frighten him. There are usually a few folks who have experience watching Flaco at each fly-out. Figure out who they are and follow their lead.

It's hard to argue with this advice. Certainly I've felt the same way on many occasions, including during my winter of the snowy owls. At West Dennis Beach you didn't even have to look for the owl when you got to the parking lot: you just looked for the line of cars of people there to see the owl. There were times when the crush was too much, when the snowy owl landed in a tree near the parking lot, for instance, and hordes of photographers, each armed with thousand-dollar telephoto lenses, tried to muscle each other out, not content when the photogenic owl—remarkably tolerant when not chewing on a duck or other prey—allowed them to get as close as fifteen feet, but finally flew away when they insisted on five.

And, unlike Flaco, snowys are not nocturnal and were not trying to catch up on sleep during the day. Which makes the Flaco-watching potentially more troubling.

And yet poor behavior was an exception. Actually most of the people I encountered watching snowys had been as delighted and as respectful as the woman who saw her first wild owl back on Coastguard Beach. I think it was important that you had to pay a price to watch the birds that winter. While they were fairly easy to find, watching them for any amount of time was something that had to be earned. That was because, along with the owls, freezing temperatures had also traveled down from the Arctic.

As a reporter, I have tried to remain agnostic in this debate. But while I am sure that there were bad apples, not to mention simply ignorant apples, among the Flaco followers, the photos I have seen and the many accounts I have read show a core dedicated group that seemed not just respectful but joyful about what they were seeing. While I recognize in Bruce my own frustrations with the crowding, the insensitivity to the bird itself, and our national tendency to turn everything into a media circus, I also recognize my own foibles and flaws. From the very beginning the writing that accompanied Bruce's beautiful films was not just moralistic, but territorial. Bruce is extremely knowledgeable and well spoken, but there was also something proprietorial about his attitude toward Flaco-watching. It was his way or the highway. Along with an admirable protectiveness for the owl, a sense of ownership comes through. It is okay for me to watch Flaco because I do it the *right way*. At times when he scolds the uninitiated on his blog you can practically hear Ted Knight yelling, "This sort of behavior will not be tolerated at Bushwood Country Club!"

It is clear that Bruce, as much as any of the other watchers, delighted in Flaco's almost daily growth and felt a deep affection for him. But the hypocrisy in Bruce's stance is also clear. He doesn't want people following Flaco, but he gets to follow him as much as he wants. And he also obviously benefits from the X posts of where the bird is.

David Barrett wrote of this underlying hypocrisy in *A Big Manhattan Year* long before he had heard of Flaco or Bruce Yolton: "You might think that those opposed to public alerts of owls would, in principle, stay away when the word leaks out, as it often does, of an owl in the park. Ha! Of course not—they go out and see the owl. Then they call or text their friends to come and see it, so it is a matter of who is in 'club owl.'"

And yet, while acknowledging the hypocrisy, it is hard to argue with Bruce's main point. Crowds, especially thoughtless crowds, no doubt stressed Flaco.

But he sure can be ornery in his blog. He grows particularly incensed by those who see the Flaco story as a parable of "freedom." From the beginning he is rooting for the Wildlife Conservation Society to capture Flaco, and when others change their minds on this, once Flaco proves he can feed himself, he remains a staunch proponent of capturing Flaco and remains so right up to the end (while also being clearly fascinated by watching the owl grow and adapt to his new life). Those who don't agree with him are anthropomorphizing softies who just don't understand. He rails about their "bias against captivity," a strange phrase if there ever was one. He deplores the vandals (they are always "vandals," a favorite word) who freed Flaco and sensibly suggests that if Flaco were recaptured he could be given sanctuary in a much larger space than his previous cage. But there is a lapse of logic here. If the deplored vandals had not cut the wires of Flaco's cage in the first place, the poor owl would have spent the rest of his life in his department store window of a home.

Here is Bruce in February, not long after Flaco's escape:

> I'm disappointed to learn of the decision of the Wildlife Conservation Society to leave Flaco out in Central Park for the time being. I would have liked to have seen a different outcome. While it's been wonderful to see that he has adapted to the park, and managing well, I don't think it is good for the other birds of the park. I also worry that life as a celebrity bird isn't any better than being in the zoo. On Saturday, rather than let him roost quietly, there was always a large, noisy crowd

watching him. He hooted while I was there in the afternoon, and he didn't seem to be getting much sleep. I find it frustrating that [what] could have been a teaching moment to let folks know about how to report and watch owls has been lost. He's now being promoted as the "Next Big Thing" and a "tourist attraction."

And here he is a couple months later in April:

To most New Yorker's [*sic*] and the media, this was a wonderful tale of how a bird having lived in captivity could become "free" and thrive in the Big Apple. Unfortunately, this over simplifies what has happened. I have a number of concerns that I wish would be discussed in the media and by the fans of Flaco. My first concern is the normalization of discussing the whereabouts of a roosting owl and the anthropomorphizing of a feral bird. Flaco's location is being publicized daily on Twitter. While you can make a case that "he is used to people" and that having crowds come see him doesn't bother him, it seems hypocritical to celebrate that "he's now a wild bird" but then treat him as though he's still in a zoo. It also is teaching new birders that it is fine to broadcast on social media where an owl is roosting or nesting. I know of no place other than Manhattan where this behavior is tolerated by the local birding community. I also am concerned that the "groupthink" that he's now happier prevents any scientific rather than emotional evaluation of what is best for Flaco and the other wildlife in his current territory.

And so Bruce goes on through the year:

May 9, 2023: Flaco continues to hunt the Brown Rats that surround three dumpsters at the compost heap. His situation doesn't seem to be the idealized wild experience you read about on Twitter. He's not living a wild life. He's a feral bird just getting by in a city park.

June 6, 2023: Last night I watched Flaco hunt at the compost heap. While the photographs of him look wonderful, the environment is far from it. He's hunting at three dumpsters in the middle of a staging site for the construction of the new rink/pool. I'm pointing this out because on social media, Flaco's freedom continues to be celebrated as though he is living a full life. He's living a different life than when he was in the zoo, but it's now at a dump site, in an urban park. Far from a full life.

August 14, 2023: Watching owls takes patience, but most importantly it requires you to do your best to stay out of the owl's way. After six months, I wouldn't say Flaco trusts me, but he does seem to know he can ignore me!

September 1, 2023: He ended up having folks disrupt him multiple times while trying to eat and once had to leave his prey. He waited 45 minutes before returning for it. Even after midnight, he hadn't been able to eat much of it.

PURIST OR POPULIST? I suppose, like most people, I have a foot in both camps.

Leave the poor owl alone.

But also: *I want to see it.*

As a card-carrying Thoreauvian (I mean this literally, I belong to the Thoreau Society), I have a lot of Bruce in me. "I have never wasted a walk on another," Thoreau said, or something close. Whatever his exact words, I have always preferred seeing my nature in solitude. I shy away from groups, abhor the idea of tours and tourism. In theory at least. Of course I'm not unaware of the world we live in, a world of concrete and constant dings to distract us. A world where the limited wild is often the only kind available. Which is to say a world where you better take your wild where you can find it. Where you might be able to retreat to a cabin in the woods (if you can afford it and find it on Airbnb) but you won't ever really get away from it all because *it all* includes those pesky electronic signals that beam over almost every inch of the globe.

As well as tours, I don't like cute animal videos or zoos or even nature shows. I fully understand the desire to give an animal its space. And I get the irony of those who rail against the zoo but who unintentionally create zoo-like conditions, dozens of people gawking at the bird.

But a huge and beautiful owl in the middle of the city. Who can resist that? Who wouldn't want to at least get a glimpse?

I am, as always, of two minds.

The truth is I'm a lapsed Thoreauvian, and if it is raw nature and wildness I'm after, I've come to the wrong story. After all Bruce is right: Flaco was overchronicled, overfilmed, overwatched. That was one of the complaints from the very beginning: the crowding, the overfilming, the voyeurism, the overexposure.

A celebrity bird.

An internet bird.

For many people, a virtual bird.

ON THE NIGHT OF May 1, three months after Flaco's death, I host a gathering. Not around a campfire or in my living room, but on my screen.

All in all about twenty people take my Zoom invitation on X. These are Flaco followers but with a difference. Some had actually seen Flaco in the park, but most only saw him on their computers. For thousands of people, maybe millions, Flaco was a flickering image made up of pixels on a screen. And yet many of these people didn't see following him in this way as a thrice-removed experience. In fact, many of these people claim to have fallen in love with him.

I wonder: What if we take these people at their word?

I, for one, am not going to deny them what they feel. In fact why not let them speak for themselves?

Laura from Boulder, Colorado, who works in wildlife rehab and rescue, begins by speaking of the "enormous emotional aspect" of following Flaco every day, and Elizabeth from San Francisco talks about the painting of Flaco she completed on the one-year anniversary of his escape. "I started following him right from the beginning," she says. "I was about to quit Twitter but started to see pictures of Flaco and I couldn't quit. I checked in every day with whatever was happening. The beautiful pictures and videos just made me feel good. And I felt part of a community."

C. J., who is in New York, downtown with her mom, and who went looking once in the park but never saw Flaco "in person," still says she "feels like I'm in mourning." She does not define herself as a birder and is clearly a one-bird woman: "I can't look at pictures of other birds now. Anytime anyone posts a picture I think, 'I don't want to see it.'" Her mother confirms this. C. J. admits to being mildly obsessed: "I couldn't believe more people didn't know about him. Every time I got in an elevator with people I would start to tell them about Flaco." She has made a shrine in her house as a tribute for Flaco, adorning a vase with a little branch and an owl made out of wood and hay. "Flaco took me on a big, kind of eye-opening emotional journey. In the beginning I was so worried about him. I saw him as an icon for New York, resilient and resourceful, everything we think about ourselves as New Yorkers. But gradually I changed. I understood that these were my presumptions. I didn't really know how he felt. People were calling him wild and free, but he could have been terrified. He might not feel free. He might be afraid every single minute. I wish I had not imposed my own ideas on him. I wish I had done more to help him. Anyway, he took me on this really big ride."

Sandra, the woman who told me about feeling the wind of Flaco's wings when we met in the park, talks about her conflicting desires to always go see Flaco and the need to restrain herself: "You have to be aware of how many people are looking at him at any time. You need to think about it. 'Should I really go see him right now?' Because we all need our space. Who knows if he felt crowded?"

Cora's voice is quieter, heavily accented, tentative at first but gaining in power as she goes on. "I work in a hospital, very close to the park, and one day I walked with a doctor to see him on my lunch break, and after we kept telling our coworkers, 'Come see this owl because he is a special owl who visited us in Central Park.'"

Her real encounter led to many virtual ones, but they all felt real enough: "After that I followed him on Twitter and Instagram. I gained friends. It's like meeting under a tree and asking, 'Where is he?' Is he in the elm tree or over by the compost? We get to be alerted every day. 'What is he doing now?' we ask each other. It's kind of magical and gives me hope. It's peaceful when you're under the tree. It's comforting."

A few days after our Zoom event, I will ask Justine, my graduate research assistant who handled the technical aspects of the call, to write up her own short summary of the conversation. She will write:

> The thing that struck me from the very start of the meeting was the way every person was emphasizing, in their own way, the sense of community they felt around Flaco. Regardless of where they were and whether they were following him in-person or online, there seemed to be this sense of connection and joy that was spanning across a lot of divides. In some ways, it felt like a movement in opposition to the increasing calcification and disconnect that is present in our society these days—and I was really intrigued by the potency of one bird to do that. I also found it amusing how fascinated people were with the idea of freedom—how freedom

was viewed as inherently heroic, and I was wondering how much of the allure of that freedom comes from the fact that we generally experience very little "wild freedom" anymore in our own lives (speaking for most people), and to what degree Flaco was taking on people's projected desires and unmet needs.

This theme of connection pervades the Zoom call: Dave, aka @bangkokdave, says:

"Flaco's story was a way to connect with something more wild, but not fully wild. That was an important element. He was neither fully wild or really captive. He was straddling the two worlds, in-between. He was kind of here and there. Some moments he was Flaco the captive and some he was really a wild bird. One talon in one world, one in another."

And:

"I think one of the reasons we connected with him was that he connected with us. I think that's because he spent so many years in captivity. He must have bonded with some of the zoo workers. Several photographers who saw him regularly said it looked like he was really watching the crowd. I think there was a real communication going on from him to us. In various ways, silently, I think an unconscious connection existed between him and us."

Another woman, whose name I don't catch, responds: "Nan Knighton fell in love with him. But what I wonder: Was Flaco in love with her?"

"Maybe with her air conditioner," Sandra says.

But then she adds more seriously, "He knew the difference between people."

I tell her that that is something all the photographers I have talked to agree upon.

Someone comments that during his city sojourns he looked into people's homes as if "reversing what had happened to him for thirteen years."

Sandra says, "He kind of took on this superhero trait. Whose window is he going to grace next? Who's going to be the lucky one? Who will Super Flaco grace?"

Elizabeth adds: "He was beautiful and healthy, but I always felt a little sad because he was not going to find his mate. He was alone and had no one to hoot with. Sometimes I hear great horned owls and they hoot and then another one hoots back and then they hoot together. And it's beautiful. So I always felt a little sad."

A doctor from New York named Afshan Khan joins us and says she wants to donate a gallery space and host a live Flaco event: "I love the bittersweet story of Flaco. He is my hero."

Someone brings up the use of rat poison, and someone else says, "Flaco was facing the same threats that all wild birds face." Dr. Khan replies: "It's a real health hazard for those of us who live here, for tourists, for dogs and cats, too. The poisoned rats die and it all seeps down into the ground and comes into our water supply. There is no reason in this day and age to kill them so inhumanely."

As we start winding down, the talk turns to what might have been different. You might think a crowd like this wanted nothing but freedom for Flaco, but now, their ideas perhaps softened in retrospect, most say that they were not against some sort of salutary interference. There seems to be a general sense that ultimately Flaco should have been humanely caught and moved to a kind of sanctuary. Where? Some suggest Europe, where Eurasian eagle-owls fly free; another person nominates the Adirondacks. If this sounds far-fetched, it is not too far from the ideas that purists like Alan Drogin or even Bruce Yolton have suggested.

"I wonder if we could have saved enough money to get him to Germany or Madrid. To where there are other owls."

"I always wanted to move him to a sanctuary."

When the talk turns to the zoo's role there is frustration and anger. I am impressed by the level of emotion.

"The zoo failed him both in freedom and captivity. They probably could have saved him. Once they decided not to catch him, they said they would 'continue to monitor him.' But they never made an effort to actually track him. How did they monitor him? Was there any evidence of this?"

I have been reading up on zoo ethics. I understand that the World Conservation Society is not a bad organization; in fact much of what they believe is in line with my own environmental philosophy. It seems to me that in the case of Flaco their main failure was that of narrative. By saying nothing, they let others fill the void. While it is not clear what they could have said to appease the masses, saying nothing did not work out well.

Certainly not for this group.

"There was never a plan. There was never a safe sanctuary. A safe haven somewhere. They never went out of their way to do this. Something like that could have saved him."

"They simply failed. A failure of imagination. A failure of planning. Just failure to seize the moment. They had this incredible moment that they could have used to help him."

"I grew up across the street from the San Diego Zoo. Different zoos are very different. They have enormous cages for birds there. There was an opportunity to take responsibility, to rise to the challenge and cast zoos in a different light. And to say, 'He's not ours anymore, but we still care about him as a sentient being and we'll do what we can for him.'"

"They abdicated responsibility. After February they drop out of the story entirely."

Near the end of the call we make plans for a Flaco celebration in the gallery space the doctor has offered. We agree that it should be on Saturday, the day after Solstice.

One by one the attendees blink off. It isn't until they are all gone that I take the time to read the chat. This is the first comment:

> Hi, I'm Rochelle. Am terminally ill and at this moment have a cold, so, not wanting to share my picture. Live in Los Angeles. Daughter has been an avid birder since second grade. Since she lives in NYC, she started sending me pictures of Flaco and I "followed" various people.... Flaco has given me a measure of happiness during this incredibly difficult stage of my life. I am so grateful that such a wonderful creature was alive and I could experience him.

I SHOULD MENTION THAT while I am doing most of my reporting for this book in New York City, I'm doing most of the writing back here at my desk in North Carolina. One of the habitats I've been getting to know pretty well over the last couple of months is LaGuardia.

Unlike Nan Knighton, David Barrett, and David Lei, and so many of the Flaco-watchers, I'm no night owl. Unless you count the time when I get up each morning as night. I've found that rising at 4:00 and getting to work by 5:00 is a good way to carve out some quiet time and get my writing done before the demands of the day take over. But it hasn't been all that quiet the last few weeks. While I type I am being serenaded, that is, being hooted to. It is not Flaco's hooting—single deep hoots—but rather sometimes three couplets—*hoo-hoot, hoo-hoot, hoo-hoot*, followed by one more emphatic final note—*hoo-WHO*. The call comes from a barred owl, sitting in a tree not far from my window, and its calls are varied. Sometimes I hear the call it is most famous for: *Who cooks for you?* But that's the least of it. As the American Bird Conservancy writes: "Other calls include hoo-hoo, hoo-WAAAHH, and hoo-WAAAHHH calls, used in courtship. Mated barred owl pairs often duet, resulting in a cacophony of weird howls, yelps, and raucous squalls, which can be unnerving if the listener doesn't know the source!"

A few times I've tried to track the owl down. One day, in my excitement, I set off the home alarm, waking the whole house. Another day I searched futilely in the trees, and another I pushed off in my kayak when I heard its mate hooting across the creek.

Yesterday I had my closest encounter yet. I had just poured my coffee and settled in to write. It was a little after five when I headed out to our cul-de-sac, and the first animals I encountered were not owls. Our neighbor has two dogs, squat little crazy beasts of indeterminate breeds, bustling and chesty, both with Napoleon complexes. They like to either run up to you and greet you happily or start barking wildly while looking ready to bite. They showed their friendly side at first and trotted next to me as I went owl hunting. I reached a tree that seemed to be vibrating with hooting. I stared up. It had to be right up there. It was exciting to be encountering a real owl, not a virtual one. But at just that moment I had moved far away enough from my house and my neighbor's house for the dogs to decide that I was foe, not friend. Suddenly I was a stranger. They started barking wildly, loud enough to wake the neighborhood and, it turned out, to flush the owl. There it flew, down from the tree I'd been staring at, across the cul-de-sac, and off toward the creek, a dark swooping vision.

I was irked at the dogs for barking, and at myself for just the kind of behavior I (and of course Bruce) was critical of, getting too close and disturbing the bird. But there was also this: the barking had flushed the bird and so provided me with my first real sight of the owl, a clear if brief view of the raptor in flight. There are better ways to earn such a vision, but things are, as they always are, complicated.

The same can be said of my snowy winter on Cape Cod. For me the story of the snowy owl irruption was a story of wonder and wildness. This was partly because I made a conscious decision to just experience the story, not *cover* it. During the first days of my own owl quest, I decided *not* to read up on the subject. This was my attempt to have more of an unpremeditated experience, to perhaps re-create the spontaneity and wildness of that first unexpected owl encounter I'd had long ago on the little beach. I wanted to see owl, experience owl, maybe even if I was lucky *be* owl for a second or two, before I read owl.

Or maybe the reason I didn't bone up on the bird at first is that I have read and written enough environmental stories in recent years to know where they all end up heading. How nice to have a hopeful story, a story of humans rushing out to take joy in encountering other animals, a story that didn't end with the doom of the planet brought about, as always, by the evil done by us. The modern environmental mind naturally and understandably runs toward self-castigation. Isn't it nice to have a story that doesn't?

So I felt a little sag in my chest later when I finally started doing research on snowy owls and discovered that our old friend climate change was being dragged out as a possible culprit. I learned that the boom of populations of lemmings was what led to the boom of owls that, in turn, drove them south looking for new territory. My friend Scott Weidensaul assures me that this is not (for once!) something that is climate driven. But if you can't say climate change is directly behind the mystery of the irruptions, warmer temps in the Arctic can't be a happy thing for the owls in the long term. And indirectly? As the popular eBird site puts it, "These owls are surely telling us something, but we still don't know or understand exactly what." It will not be terribly surprising if the answer to that mystery is an ugly one. John Schwartz, writing in the *New York Times*, quotes Cornell ornithologist Kevin J. McGowan on the disturbance of the snowy's arctic environment: "That has to be one of the most vulnerable ecosystems on the planet. That's going to be one of the first places that falls apart when there is warming in the atmosphere."

It may turn out that the arrival of these radiant visitors is yet another dark symbol of the end of the world, or at least the end of the world as we know it. But for those weeks on Cape Cod I chose to think otherwise. Each morning I bundled up and got out into the cold with the owls, keeping my distance while watching and living for just a moment in a world other than human, while choosing to see their arrival on the Cape not as the tolling of the usual bells of doom, but as a generous visitation, an unexpected joy, an undeserved but well-appreciated gift.

7

THE PASSION OF ANKE

IN HIS BRILLIANT manifesto *The Abstract Wild*, Jack Turner mourns the loss of our deep connection with animals. Implied in this is our unwillingness to fight for wild animals. He writes: "To reverse this situation we must become so intimate with wild animals, with plants and places, that we answer to their destruction from the gut. Like when we discover the landlady strangling our cat."

Turner has a particular disdain for zoos. When he was sixteen, he saw the tan streak of a mountain lion while hunting along the edge of the Marine Corps Base Camp Pendleton in Southern California. He could never shake the sight of that lion, and many years later he found himself staring through the bars of a small cage at an imprisoned mountain lion in Mysore, India. Another spectator began to taunt the animal by throwing "a kind of popped grain" at it. The lion just glared. Turner couldn't handle the indignity. Before he could think, he attacked the man. He writes: "When I reflect on that day, I produce reasons for my behavior that my culture says do not compute: the lion had no freedom, she was far from home, she was being humiliated." He had acted without thought, lashing out not just at the taunting man but at the zoo and "the son of a bitch who had trapped the mountain lion and shipped her to a living death of unending humiliation in a squalid hole in Mysore."

We have words for the way he acted: *immature*, *irresponsible*, *criminal*. But for a moment he was driven by something beyond the everyday, something that many of us feel but almost no one acts on. Zoos are not inherently bad places, perhaps, and some are much better than others. However, they are, Turner believes, always about control.

So many of our relationships with animals are about control. Consider something as seemingly benign as birdwatching. What is it we are trying to experience when we go in search of birds? Our motives are, as always, mixed. We come to them as watchers, as hunters, as collectors, as acquirers. We come to them to check an item off our list, to fill ourselves with a sense of accomplishment and pride. But also, if we are lucky, to briefly leave ourselves behind. We also come to them for aesthetic pleasure, watching the colors of a painted bunting, or as sports fans, cheering on northern gannets as they plunge from a hundred feet up. We come to them because we are *getting something* from them.

Scientists, our society's high priests, might seem beyond this. Far from it, Turner tells us. Scientists are in fact the ultimate controllers, justifying this by the pursuit of knowledge. They tag and drug and measure animals, and we accept it all as a matter of course, as necessary. "Knowledge and control are indivisible," Turner writes. He makes a radical suggestion: "What if the effect of scientific experts creating environments, treating ecosystems, and managing species is (sometimes, often, always) as bad, or worse, than the effects of unmanaged nature?" And: "The great need is to imagine an alternative. Perhaps we don't need more information; maybe the emphasis on biological inventories, species recovery, surveillance, and monitoring is a further step in the wrong direction." Blasphemy, and never more so than in the age of climate disaster: we have fucked up the planet and so we must fix it. It's up to *us*! Wildness has no role in this! Turner knows that his ideas will be pushed briskly aside, regarded as naive. The Earth must be controlled. But, he counters, consider how we, how all life, evolved and can continue to evolve.

"Life evolves at the edge of chaos, the area of maximum vitality and change."

I AM THINKING ABOUT a different possible relationship with animals. I am thinking about another way to watch Flaco. I am also thinking about obsession. I am thinking about Anke.

The first time I talk to Anke Frohlich it is by phone, and I have no way to record it. This I will deeply regret, as I soon feel like I am talking to Werner Herzog or maybe more like one of Herzog's characters. It isn't just the German accent. It's the intensity, the almost anachronistic passion for a thing. We will talk again, but as with so many things the first time is not recoverable. I am relatively low tech, and when I do phone interviews, I call from my wife's phone, put it on speaker, and record with my phone. I'm sure there is an app for this, but whatever. My way usually works, only this time my wife is away, so I am scribbling madly on a pad as Anke talks, my hand moving fast enough to create an instantaneous case of carpal tunnel.

Anke is the most extreme of Flaco followers. One of the core group of Flaco's late-night tribe, she would often stay at the park until it closed at one in the morning. Early on, she says, she made peace with the danger, and while others sometimes carried means of self-defense, she only carried her camera equipment. That equipment was heavy and awkward, a tripod and an enormous lens for night shooting, and she lugged it onto the subway near her apartment in Greenwich Village and headed north, a nearly three-hour commute back and forth almost every day, so she could put in what would amount to more than a thousand hours of Flaco-watching or sometimes, just Flaco-searching.

"Some nights I walked for a very long time, for hours, finding Flaco or not finding Flaco so it was a lot. I mean it was actually bone breaking at times and I have to say it took a real toll on my body, and yet there was also this excitement and this, just this, *I've got to go*. I felt so drawn to it; it became such a passion that I couldn't not go, and one night when I decided not to go because my body was hurting, I missed out on a fantastic night, which then I said, 'Never again, I'm never not going again.'"

Like so many people, Anke turned to bird photography during COVID. She developed a deep attraction to raptors that had led to an almost prescient trip back to Germany to look for, of all things, Eurasian eagle-owls. This was B.F., before Flaco, and she had searched the quarries of her native country, hoping to take a picture of an eagle-owl in the wild, but sadly couldn't find one. And then when she got home to New York, her wish was granted when an eagle-owl appeared in her city, "at my door."

Is it an exaggeration to say that one of her life's bigger regrets was not going out that first night that Flaco escaped? I don't think so. She was exhausted after driving back from photographing short-eared owls in upstate New York when she saw the first tweets about an unusual owl on Fifth Avenue. Anke was one of the first to recognize it for the species it was, the very species she had gone to Germany to find. But then she saw the sad picture of the owl next to the cat carrier and assumed he had been captured. To this day she resents the fact that that night a few people privately messaged each other, including David Barrett and David Lei and Bruce Yolton, and that she was not in the loop.

"But the next morning, bright and early, I looked at Twitter, and nowhere did I see that he had been captured. So I grabbed my camera and I thought I'll just go look for him, and when I did, I found him immediately.

"I have the picture from that first morning that I'll show you. He was hanging really high in this barren tree in front of the skyscrapers with the Essex sign in the background. It was absolutely frigid, one of the two coldest days of the whole year. And I sat down on the frozen ground to stabilize my lens because I didn't bring a tripod. The crowd had already started to grow. It was freezing and I'm in a winter coat sitting on the ground trying to get my first pictures of Flaco, and not long after that the first reporter came and stuck their microphone in my face and said, 'Would you like to comment?'

"And I said, 'I just arrived here. I'm looking at the same thing you are. You're looking at this beautiful owl. How do *you* feel about it? Incredible, right?' And then I went home and got my big lens and the tripod and came back and spent the whole day watching him and the next day again. So those were the first two days and I was hooked immediately. Pretty quickly I began to have this inner need. This need to watch him grow. To be part of this development. To be with this animal who had been deprived of the world. To watch him learn. It was a unique experience. In the end it was a very personal experience."

"I fell so in love with him immediately. He was so beautiful. And I thought, 'I hope he makes it.' It got so I didn't like not being with him. The second I got home I wanted to go back. I wondered what he was doing."

Late at night the Flaco crowds vanished. Then only a core group that was sometimes very small, four or five, sometimes smaller, sometimes just her and David Lei, remained. David Barrett had usually returned home by then to write up and post what he had seen at flyout hours earlier. Sometimes Anke was alone. She didn't mind. And she had always been a night person; it was when her mind was clearest.

"You have no idea where he is at first and you are looking for him in the dark. Your senses are open and you are listening and looking. The hunting is part of the appeal. You are outdoors in the dark. It stills the mind. It makes you feel alive. And that feeling of being in love gives you energy, too."

She was not sure she should be in the park at night, but she committed to it, despite everything, despite good sense, you could say. "I decided to take the risk, to trust the experience."

Anke was basically exhausted for the whole Flaco year. She would get home late, but as soon as she got home she would want to look through her photos and see what she had captured. "I would stay up until three or four going through my pictures because I just needed to look at what I got," she said. Since she took the pictures in the dark, they would reveal surprising things, things you would never see with the naked eye. The way Flaco tilted his head maybe, or a new expression, or where he perched. Looking at the film became part of the process of discovery.

There wasn't a lot of sleep during those months, but the excitement of it all, the obsession of it all, kept her going.

"When you get obsessed with something you have this energy," she said. "And it was always exciting. We all grew with Flaco."

I mention how David Barrett said he did not know what to do with his nights after Flaco died.

"I understand that, believe me. I was lost at night for a while. I guess I still am. I miss him dearly."

DURING MY YEAR RESEARCHING and writing my osprey book I spent most of my time focused on the birds. There was a point, which I wrote about in the book, when Nina came out on our back deck and found me ripping apart a fish from the market with some pliers. I was wearing just a bathing suit, but it might as well have been a loincloth. She gave me a look that was a combination of *What now?* and *This has gone far enough*. I explained that I wanted to experience what it felt like for ospreys to tear into the skin of a fish with their bills, but oddly this did not placate her.

Jack Turner suggested we need to become intimate with animals, but I am somewhat pessimistic about our ability to really know and empathize with other species. With so many other things crushing in on human beings in our crowded lives and all the other self-centric concerns and the difficulty of extending our imaginations beyond even our immediate families to other human beings, how is it really possible to care, to *really* care, about a member of another species? The only answer appears to be intimacy. But how to achieve the intimacy that you have with your dog or house cat when you are dealing with a cougar or an osprey or an owl? The history of the last three centuries can be described in many ways, but one way is the age of separation of *Homo sapiens* from the rest of the animal world. Think of the amount of time (millions of years) that humans and humanlike creatures spent in close contact with wild animals as hunter-gatherers, and think of the relatively miniscule amount of time since. It is a new world for us, and in this new world we are all alone. Physically separated, it makes sense that we would also separate ourselves emotionally.

I do not claim to be an exception to this. My wife, for one, has a much greater natural empathy with animals, especially the ones who live in our house. My primary interaction with wild animals is as a bird-watcher, and as the name implies there is often something cold and removed about that activity, watching from outside. My occasional moments of empathy with birds may not be empathy at all but rather an emptying of thought, a kind of thoughtless delight, and a delight not in our commonality but in their otherness. On the other hand, the time in my life when I had most empathy for birds was during the year I spent writing my osprey book. Every day I went out to the marsh to watch the nests, and soon I was rooting for the young to fly or worrying one of the parents had perished. There was a simple reason for this: ospreys were what I did all day.

I think of Alan Poole, who wrote the definitive book on ospreys and their behavior. Some years ago now Nina and I rented his house in South Dartmouth, Massachusetts, a small cabin tucked into the trees that overlooked a tidal creek, an osprey nest, and the ocean beyond. While staying there, I found an old moldering notebook in the shed, one from the early 1980s when Alan conducted his osprey studies. It was a mouse-eaten steno book, and a typical entry might read:

7:25 Male returns to nest with female
7:37 Male leaves nest
8:20 Male returns with fish

And so on. Simple factual entries but not without some spare poetry. And, when mixed together with a thousand other similar entries, they add up to a picture of the birds' world that in the end gave life to the book he wrote. The field notes were the bricks with which he built his knowledge of ospreys, but somewhere along the way he also fell for the birds. It makes me wonder whether the best science, for all its pretensions to objectivity, doesn't require a dash or more of love.

My own best writing about ospreys, I'm sure, was filled not just with admiration but love for the birds.

At times I worry that my relationship with ospreys, or with dolphins or owls for that matter, is sometimes not so different than my relationship with a beautiful painting: I am briefly lifted out of myself—I *oohh* and *ahh*—and then I go back to my daily business.

Years after I wrote my osprey book an astute reader came up to me at a reading and told me that my book was not really about ospreys, that it wasn't really ospreys I was obsessed with. "Oh," I asked, "what is it *really* about?"

"Writing," she said.

I nodded. She was at least half right. I had learned a whole lot about ospreys but as much time as I spent observing them the moments when I felt most alive and absorbed and exhilarated were when I was recounting what I had seen. If the birds spent most of their days fishing, I spent mine typing.

THE NEXT TIME I talk to Anke is in her apartment not too far from Hadley's dorm. She looks the part I already cast for her during our phone call. Taller than me, perhaps six feet, with a long Valkyrie blond braid running down her back. Her apartment is neat and well lit, and mourning doves gather at the window. She reaches through a hole in the screen and feeds them organic sunflower seeds. (Blue jays, she has learned, prefer organic almonds.) We sit on the couch and talk Flaco. I can't imagine we will possibly approach the intensity of our phone call, but we get close.

There were times when her Flaco encounters edged toward the shamanistic. Back when I studied ospreys I heard the story of primitive hunters in Bolivia who had sliced open their arms and slipped osprey bones under their skin to enhance their hunting skills. That was roughly the level of intensity that Anke had brought to her Flaco pursuit. Not that I actually believe she had owl bones under her flesh. (I forgot to ask if she has a Flaco tattoo. I wonder how many depictions of Flaco have been inked into the skin of his various followers during the past fourteen months.)

Like a couple of the participants in our Zoom call, Anke sometimes thought that Flaco was communicating with her. Certainly he recognized her after all those months.

"He trusted me. One day he landed very close to me. We were both surprised."

That day their interspecies connection was not mystical but practical.

"He just stared at me and looked at the ground where I was sitting. It took a minute for me to get what he was telling me. He kept staring and I finally understood: he was telling me to move."

She complied and sure enough Flaco immediately moved in and started hunting on the ground where Anke had been sitting.

It is important to remember that Anke was rarely just *looking* at Flaco. She was also taking photographs of him. In other words, there was often a lens between them.

"I was aware of this from the beginning," she says. "That the camera could take away from the bird. That it could get in the way of the experience."

The camera complicated things.

"When I first picked up a camera, I already had the thought, *Be careful.* This could really take away from your experience of being with the bird, understanding the bird, experiencing the bird, getting the richness of the experience. It could take that away, and I knew that. Yes, and I thought to myself, *Boy, I also want to take the pictures because I get a lot of joy out of seeing them later, looking at things that I would have missed because it was so fast.* And with my camera I see a lot of different things that I wouldn't have seen. Especially with Flaco at night when I would capture what nobody could see with their bare eyes. So I had that thought when I first picked up the camera and I thought, *Boy, I really hope I can strike a good balance and not miss out on either one of those things.*"

I tell her I understand. Some mornings back at home, when I hear the barred owl call, I am torn between going down and being with the owl and staying up at my desk and writing about owls.

She tells me a story from when she was first starting to photograph birds. She was on grass, really low, lying on her stomach, getting "nice shots of being at eye level" with a yellow-rumped warbler, trying to look from his same perspective, "not shooting down or any of that."

"And then I sat down on the grass and I looked at my pictures. I was very focused. I'm deleting, I'm looking. All of a sudden something touches me on the leg. I look up. There he is. The warbler. A wild migratory bird sat here looking at me, giving me the clue. *What are you doing? I'm right here. Why are you looking at your camera? I'm right here.* Exactly. And we looked at each other. I'm like, 'Oh my God, thank you. I forgot. Thank you.' It was, yeah, these moments are just like, oh my God."

Some might roll their eyes at Anke's extremism. A kind of Germanic romanticism left over from earlier centuries. And free use of the word *love*, which would give scientists the willies. But she is speaking to things other Flaco-watchers have not approached.

The Abstract Wild, more than any field guide, has become my bible as I explore the world of Flaco. Now, for Anke, I paraphrase a quote from Schopenhauer that Turner uses in the book: "In the mind of the man who is filled with his own aims, the world appears as a beautiful landscape appears on the plan of a battlefield."

Anke nods and says, "Yes, that is the worry."

We all come to the land, in this case the landscape of Central Park, with our own purposes, our own battle plans. Some went to Flaco to garner more likes on social media, some to take beautiful pictures, some of us to take away words. We might have had purer motives too. But we all had our *uses.*

YOU CAN'T PLAN THE wild. You can't organize it and check it off a list. You can't go with a group in search of it or take a tour to it, and there are no Twitter alerts as to where it is.

Flaco was the perfect bird for our new self-conscious world. The fact that he was most often viewed through a lens of some sort seems important. Maybe that is another way to define the time we live in: the time of the lens. Isn't there always, or almost always, a lens between us and experience? I walk through Central Park and a jogger runs past talking into his phone, and I record the scene by talking into mine. This is our world.

Maybe this has always been the human world. These days the lens may be that of a camera, a computer screen, a phone, but it can also be something more old-fashioned, like, in my case, a journal. Self-consciousness is nothing new. "Life is the poem I would have writ if I were not so busy living it," said Thoreau. In this double-consciousness writers are emblematic of our time, our scribbling a sickly symptom. Always living a double life. Maybe the journal was Thoreau's iPhone. Everything he saw was digested and regurgitated like owl pellets into that damn book. Everything had a double meaning. His experience was mediated experience. But what's so bad about that? Don't we digest to feed ourselves, to provide our bodies with nutrition?

Of course in our time this has all accelerated to an almost comic degree. Everyone taking pictures of each other or themselves and filming and posting and recording (while I talk into my phone and chronicle everything I see as it is happening). People walking down the street and talking to themselves like schizophrenics. We are thrice removed, maybe thrice thrice. Wildness itself, as much as any animal, seems to have gone extinct. It has become harder and harder to find, the more removed and virtual we get.

Another lens that gets in the way of our intimacy with wild animals is our own minds. We crave the authentic, the real, the wild, but we understand how hard these are to achieve. "The biggest cage is our own minds," Ruben said.

What is this elusive thing itself that we romantics are so hungry for? Could it be simply living without the battlefield map, living the way we imagine animals do, though of course animals live with their own specific and restrictive maps, encoded maps drawn with the aim of procreation and food? Realists scoff at romantics, but what is so-called realism if not just another map?

When he was young, Turner found himself in the Maze, a red rock labyrinth in the canyonlands of southeast Utah. It was not a pleasure trip. He had earlier survived a crash landing in a small plane and was rushing to meet a companion who had been dropped off by another plane and come to help him hike out of the desert. It was already getting dark, and Turner was understandably anxious, but suddenly he came upon some figures on a cave wall:

> They were pictographs, but not the usual stick figures and crude animals I'd seen before. There were fifteen of them, painted a dark, almost indigo blue. Some were life-size, some smaller. Some were abstract, like mummies with buggy eyes and horns. Others had feet and hands.

The effect of seeing the figures was visceral.

> I froze, rigid with fear. My usual categories of alive and not alive became permeable. The painted figures stared at me, transmuted from mere stone as if by magic, and I stared back in terror.

Gradually the fear disappeared, and he became fully absorbed in the pictures. He didn't want to leave.

But years later when he finally went back with a group to show them what he had seen they were just pictographs. They elicited no magic. He said he had become "a tourist to my own experience."

Why is that? Why are we all so often a tourist to our own experiences? Why can't we get back to raw experience?

Perhaps our craving for raw experience is always at least partly nostalgia. My friend and colleague, the poet Mark Cox, has written, "So much of what we do is just an effort to duplicate some original feeling or experience that can't be had again."

Maybe that is why we love the first moments, the unexpected moments, before the precedent has been set. The moment when we aren't searching but finding or being found. When Flaco showed up at Nan's window or the very first night, before the stories started, the moment when Edmund encountered a strange owl on the Fifth Avenue sidewalk.

THE DAY IS SO celebrated in this country that it even has its own name, like Valentine's Day or Halloween. We call it Super Sunday.

I am not talking about the celebration in New Orleans in March. I am talking about our unofficial national holiday, a day for nachos, for beer, for celebrating the definitive American art of the TV commercial, for gathering with friends around the tube to watch football, and, historically, for a record number of cases of domestic violence.

It was also a day, on Sunday, February 11, 2023, when the world's cameras were pointed mostly at a different kind of Eagle, those playing the Chiefs, and not a world-famous eagle-owl.

Which meant bliss for Bruce Yolton, an evening when Central Park was quiet, almost empty. The human mobs were gone, and Bruce found Flaco "roosting in a tree further west than the White Pine" where he had been on the previous days. Flaco means *skinny* in Spanish, but there were certain poses, when his chest is puffed up and his chin tucked in, where Flaco looked the opposite of thin.

This was still pretty early on in the Flaco story, Day 11 of his freedom, but those who watched him carefully, like Bruce, had begun to see that Flaco was a bird of a hundred faces—and that afternoon he showed off at least a dozen. He had a kind of pensive look off into the distance, his eyes half closed and feminine, looking to all the world like a Maine coon cat until he heard something that caught his attention and swiveled his whole head, suddenly staring right down at you, orange eyes wide open, glaring like an angry gym teacher. That afternoon he also did a few of the usual head bobs, head push-ups—up and down with sometimes a little swirl to the side—all the while showing off his sharp black feathers that made him look like he had leopard spots. Then he suddenly seemed scared or agitated and looked slightly bedraggled, and it soon became clear why: he coughed up a pellet, opening his sharp jaws and choking up several pieces, a sort of graceful vomiting, then coughed up another before shaking his head to be sure he got it all out. After that the nearby twittering of some white-throated sparrows caught his attention, and there was some more head bobbing and swiveling. Up until that point he had been facing the cameras Bruce and others had been pointing at him, their clicking and whirring mixed with the sound of kids playing, but now he swiveled his head 135 degrees so suddenly there was no face at all, just a faceless body. Not long after, Flaco took a couple steps forward with his oversized feet and curled black talons and pushed off into the air. He circled the area and landed on a rock.

Though it was a quiet day in the park, Bruce wasn't the only one watching Flaco. As usual, Anke had taken the subway uptown, and she, too, was happy with the quiet that Super Sunday brought. The big news the previous two days, Friday and Saturday, was that Flaco had caught rats. Anke and the other photographers were among the first to see this, and there was a little debate among the in-crowd about whether or not they should tell the zoo. But by Saturday the zoo workers knew. That day they pulled their vehicle pretty much right below the tree where Flaco was clutching his latest rat, looking down at the workers, not eating it right away. Bruce believed Flaco was simply adjusting to eating in a tree for the first time. Anke's take was different. To her Flaco looked afraid, staring down at his potential capturers: "He stared down at them with a look of both pride and total bewilderment, his face saying, like, 'What's going on, I'm puzzled, why aren't you happy for me, what's wrong with humans, that they're trying to make me lose my dinner?' He didn't start to eat until they left."

That had been the same evening, according to Anke, when Flaco started to find his voice. After the zoo workers left, and he started nibbling on the rat, those watchers who were still left heard his very first timid hoots. Then they got stronger and stronger. It was thrilling to hear this until-then mostly silent bird, hooting again and again, over a hundred times, then, after pause, another round of eighty or so.

He was hooting again the night of the Super Bowl on what would end up being a special night for both the Chiefs and Anke. After his afternoon roosting in the tree, Flaco flew over to the Heckscher Ballfields to hunt, and Anke and Bruce, along with a couple other photographers, followed, keeping their distance and giving Flaco plenty of space. Afternoon became evening and evening night, the owl's time. The photographers watched as he caught a brown rat in the middle of the ballfields and stayed on the ground with it, pinning it with one foot and ripping it apart and taking his time, his head swiveling, on alert, guarding his dinner and looking around for anyone who might intrude on his feast. Finally, he gobbled it down, only the rat's tail left hanging from his mouth. Then he marched around pecking at the ground for any leftovers, before walking rather than flying off in a silly strutting almost tiptoeing walk.

As the night deepened the other photographers left until it was only Anke and Bruce. At one point they thought they had lost Flaco, so they walked around the ballfield and almost ran right into him. They didn't see him at first because they were looking down low, on the grass, since during those early days Flaco had a tendency to sit on the ballfield grass close to the fence and just watch what the rats were up to.

"He would just study their behavior because it was new to him," Anke told me. "So he could learn what was the most fortuitous moment for him to catch a rat. What he learned was that they liked to run along the fence."

So the two of them, Anke and Bruce, were naturally enough looking down at the grass near the fence when they turned the corner and almost ran right into Flaco. There he was, perched on the backstop, the first time they witnessed a behavior that would soon become familiar to them. They were so close, only nine feet away, that Anke backed up. Flaco was watching them, studying them with his huge eyes, just as he had studied the rats.

"It was startling. Like he didn't know to be afraid of people. He didn't fly off, he just looked at us, we looked at him, and it was very peaceful. And very beautiful."

She had never been that close before.

"We slowly walked back. I should have stepped back a little more since the picture I took is slightly out of focus, because we were actually that close. And Flaco still didn't move. He just was there looking at us."

The only time she had been closer to an eagle-owl was back in Germany, six months before in the pre-Flaco days. She hadn't found one in the wild, but while she was staying at her mother's she had driven two hours to the zoo where a captive eagle-owl was kept. She had called up the day before and asked, "Do you have any owls?" When they said yes, she asked, "Might I be able to hold one?"

"And the zoo guy said, 'Well, no, it's not usually part of

the program, but yeah, come, you won't be disappointed.' But by the time I came, they had already taken the owl back to the cage. And I said, 'Look, I'm the one who called. Could you possibly bring it back out?' And they brought him back for me. And I had this incredible experience with the owl. They let me hold it, and I talked to the owl and it was so beautiful. And then at that moment I had this wish, please, I want to find one in the wild."

And now six months later she had gotten her wish, or at least she had found one in the semiwild of her adopted hometown.

With Flaco gone, Bruce headed home. Anke was alone now, and the park was even quieter. On the one hand she felt just how vulnerable she was, alone in Central Park late at night. "I'm thinking, *Boy, I really am the only person in here*." On the other hand she loved it. She had always loved the night.

"Everybody is peaceful. People are asleep. I kind of feel like I own the night. My thoughts are clearer. I can be very productive. And I've always enjoyed being out at night as well. And I've never had any fear. I know there's danger and I keep my eyes and ears open, but that's part of what we said before of being alert but with a very broad awareness. You're sensing what could be there, but without fear."

Mostly she felt calm on this most magical of nights. She had never heard the park this quiet. It was then, when she assumed the night's magic was over, that she heard the hooting. So clear on that quiet night. Hooting was still new to her, and to hear it during that still night, well . . . it was hard to describe how exciting it was.

Anke followed the hoots and discovered Flaco at a small construction site marked off by cones, another new behavior that would soon become habitual: scouring the earth for food where humans had scoured before.

After that it was just Anke and the bird, the park completely quiet except for the occasional hoots.

8

WEST SIDE STORY

AS FLACO EXPLORED the urban wild of the Lower East Side, his fame grew. For his online followers there was a storybook feel to his new urban adventures, and they thrilled to his every move as he navigated the world beyond the park. And good storybooks, from *Winnie the Pooh* to *The Wind in the Willows* to *The Lord of the Rings*, need good maps. Around this time several detailed maps appeared online that traced his East Side wanderings.

Once the bird hit the city, it wasn't only the Flaco faithful who got to delight in his wanderings. He started popping up everywhere. Air-conditioning units seemed a favorite resting spot, and more than one Manhattanite was blessed with a visitation.

A woman named Robin Herbst, who lived on the Lower East Side, managed to take one of the truly stunning photos of the owl and the city on November 8 (she thinks). When I wrote her to ask how she got such a shot, she replied, "For me, it was only ever a chance encounter. . . . He landed on our living room A/C. My cat heard him hooting and so I wanted her to be able to check out the action. When I rolled up the blinds for that window, I could only see his feathery rear end. We couldn't make out much more. So we shifted over to my office where we were finally able to see him with the city lights in the background."

Flaco would continue to startle people on rooftops and ledges, sometimes looking through windows at apartment dwellers. One morning Reilly Richardson caught a glimpse of something or someone outside her window. More than slightly freaked out, she approached her curtains cautiously, worried about a peeping Tom. She peeked out and there, staring back, was an enormous bird. "He scared the you-know-what out of me," she told the *Wall Street Journal.* Flaco spent a good part of the next three days perched on her air-conditioning unit, peering in and watching her. The papers, playing up the peeping Tom angle, asked her if Flaco wanted a show. "Well, if he did, he got one," she said. She admitted to missing him when he was gone.

On November 13, after his long odyssey to the Lower East Side, Flaco began his return north to familiar territory. By November 14 he had made it to the edge of the park. That was the day he stared into Nan's window. The next night David Barrett saw him fly into the park, and on November 17 he tweeted, "Joy in Central Park today as Flaco the Eurasian Eagle-Owl has returned to reclaim his favorite oak tree, standing his ground despite visits from a hawk and some crows."

"He spent two successive days in Central Park, November 17 and 18," David told me. "He was back in his old tree, the first day, his old oak tree. So that was a huge day of celebration for people. Flaco had come home."

Flaco stayed in the black oak off East Drive near the Loch that his followers had come to think of as "his." Hundreds of people came by that day to welcome him back and, of course, take photos. He stayed in the oak until flyout, when he pushed off and flew north toward the Harlem Meer and the construction site, his old hunting grounds. At one point he crossed Fifth Avenue and was seen perching on the Terence Cardinal Cooke medical building. Many hoped he had returned to the park for good, but after perching for ten minutes he pushed off, flying east, deeper into the city. There were no reports of him being seen that night. After a day of being stared at he had flown beyond human eyes.

By this point the freedom narrative was in full force, and many people referred to Flaco as their "hero." Perhaps saying he was "free" might have been a little facile, given the obstacles and challenges he was facing as a now-urban bird. On the other hand, *free* does not mean *easy*, and while we can't ever fully understand the subjective reality of another creature, in one very real way at least he was infinitely more free than he had been in his old life. His circuitous round trip to the Lower East Side and back encompassed about twenty miles of flying, something he could never have accomplished during those early days after the zoo escape. However he did or didn't feel about it, his territory had expanded, and he was seeing a larger world.

It is worth remembering here how cramped Flaco's quarters actually were for the more than a dozen years he spent at the Central Park Zoo. The Association of Zoos and Aquariums (AZA) recommends a minimum of four hundred square feet for an eagle-owl. That is not a huge space, about the equivalent of a two-car garage, but vast in contrast to Flaco's longtime closet of a home. In fact, some of Flaco's relatives in other zoos were lucky enough to be housed in spaces many times larger than the suggested minimum, where they could sometimes hunt for rats that happened to wander into their cages. To add to the lonely picture of Flaco's life, the AZA also recommends that eagle-owls are best kept in pairs.

No wonder Nancy Garay, whose research I am drawing on here, has written, "Once freed, Flaco became the bird he always was."

So, after nine months Flaco had become, if nothing else, competent, able to feed himself and, adapting to city life, regularly seeking out quiet courtyards within buildings to rest in during the day. This also protected him from the constant irritation of other birds mobbing him, completely understandable behavior from the point of view of the smaller birds, who weren't crazy about keeping company with a giant predator with bird-crushing talons and a beak built to slice meat.

When Flaco made that glorious return to the park in November, there were high hopes among his human followers that he would settle there permanently. This was partly selfish—what better place to watch him?—but also for Flaco's sake. The park's rats, unlike the city's, were not poisoned, and while rats could of course wander in from the streets, that did make the park safer. Other dangers lurked outside the park, including vehicular strikes and building collisions.

Despite his fans' wishes, Flaco did not settle. While he continued to hunt in the park, he began to roost mostly on the Upper West Side, hooting from the tops of buildings to the point where his serenade became expected in some neighborhoods, heard by residents, dogwalkers, and tourists alike. Hunting in the park at night and resting in relatively protected West Side courtyards in the day was a pretty good system. Like a celebrity avoiding the paparazzi, he sought refuge. He could hide out from crows and cameras by day and swoop back to the park to scoop up a rat or three at dusk. As a bonus, there were rooftop pigeons to munch on, adding some nice dietary variety. A photo or two from that time captures him with a mouthful of feathers and pigeon feet protruding from his beak.

This was the period of the water towers and the beautiful West Side buildings, as if Flaco wanted to give equal time to both ends of the architectural spectrum, the archaic/practical and the novel/aesthetically pleasing. His old regulars had a hard time following him during this period, but now and then they caught up with him. They saw him poised nobly on the spires of the Beresford, or atop the beige-pink art deco building at 241 Central Park West, or just as often atop his favorite water tower, which looked patched together and barely up to its task of holding water, perched on the tower itself but also on the ladder on the side of the tower, hooting down on the city. Though he was never seen perching on the statues in front of the American

Museum of Natural History, he was heard within a block of there, and Teddy Roosevelt, whose father was one of the museum's founders and who had a barn owl for a pet that is rumored to have moved into the White House with him, would have been pleased. It is believed that Flaco may have flown as far north as 103rd, which wasn't surprising since he regularly hunted farther north than that at the Harlem Meer on the East Side.

The reasons why he no longer roosted in the park are up for debate and in fact were hotly debated. David Barrett believed the desire to find a mate and the constant harassment by other birds were key factors that drove him out of the park, and that once he left the park at the end of October he made "a pivotal discovery that changed his life." That discovery was that the courtyards for buildings were perfect for resting during the day, free of traffic, free of disturbance, free of other birds. There he could rest on fire escape railings, window sills, and air-conditioning units. He could still fly into the park to hunt, but he only spent one more night there as far as Flaco-watchers could tell. That was a rainy night in January, when, soaked through, he passed the night in a tree. "He was not a bird who could fly well in the rain," David Barrett said. "Not a duck."

Not surprisingly, Bruce Yolton has some different ideas about why Flaco abandoned the park. On the Urban Hawks site he wrote:

> After he left Central Park, social media had some crazy reasons for why he left Central Park after two days. The most incredulous was someone defending their

posting of his roost locations, by claiming that people don't bother Flaco, and he must have moved because of American Crows and Hawks. While Flaco most likely didn't leave Central Park because of people watching him, he does get annoyed by some observers and the standard owl etiquette rules should be honored. The world is full of bad actors, and it makes sense to not advertise the location of a sleeping owl.

Bruce admitted he was not absolutely certain why Flaco left the park, writing, "He's a feral bird on the wrong continent, so this is unchartered territory." Back in the fall, when Flaco first flew south, he posted this list of possible reasons for Flaco's exit:

- As the days got shorter, he may be responding to instinctual pressures to establish a territory and secure an appropriate nesting location. This naturally occurs in the late fall and early winter. (Eurasian Eagle-Owls don't build nests. They use cliffs, take over other raptor nests or can even nest on the ground.)
- He's checking for other owls, which would be both competitive males or potential mates.
- His roosting locations were beginning to lose their leaves, and were offering less cover than in the summer. He could simply have been looking for a more protected place to roost.

Since David and Bruce were two of the most dedicated followers and photographers of Flaco, there must have been times when the two of them were standing quite close to each other while staring up through their binoculars or cameras at the bird that was their joint obsession. This could not have been particularly comfortable. By that point the hostility between the two was out in the open, and in the wake of Flaco's death Bruce would write:

> After Flaco's release from the zoo, any media publicity, especially publicity that would invite the public to watch any rescue efforts would hamper his capture. However, this occurred due to a campaign by David Barrett via his X account, the Manhattan Bird Alert, to cast Flaco's release not as a misguided criminal act, but as a chance for an invasive species to live happily in New York City. David Barrett, as he did with a Mandarin Duck, Snowy Owl and a Barred Owl, has a habit of turning birds into celebrities, not for the benefit of the bird, but to promote his social media account and to get himself in the print and television media. The media got duped by David Barrett and others with Flaco.

Other less obvious fissures were widening in the Flaco community. Flaco was harder to find now, but he was occasionally seen by Upper West Side apartment dwellers, taking pictures out their windows or on their roofs. Those people often found their way to Manhattan Bird Alert and to David Barrett, and once in a while they would invite David over to their apartments to photograph Flaco while he was resting. When he got these invitations, he would sometimes bring along a few special guests like David Lei and Jacqueline Emery.

Some of those who never made these guest lists, like Anke, felt resentful.

As for David, he was not without a little resentment about those last days. The last time he saw Flaco was on February 16, on West 86th Street near Central Park. He spotted him on his favorite water tower and then again on one of his favorite building tops, where there was an exhaust cage at the very top of it on the roof. That was where he heard Flaco hoot for the last time.

According to David, people he knew—"Reliable people, followers"—saw him again on the 17th but didn't get any photos of him.

"He was also heard on the morning of the 18th on a favorite water tower and recorded by a follower, a friendly follower of mine, at around 3 a.m. on the 18th. I went out searching for him after that and so did other people at his usual places, and we couldn't find him on those successive days beginning the 19th and following. So we didn't know what was going on with Flaco. No one saw him or heard him after the night of the 19th, or that's what I believed at the time."

He later found out that there was evidence of Flaco on the 19th.

Someone had taken a picture that night but did not post it for over a week.

"I'm not sure what this person was up to or why. I think she should have been in touch with me about having Flaco, but she lied about not having Flaco when in fact she did have Flaco, and that's bad because I wonder what she knew about Flaco in his last days."

Jealousies, deception, exclusion, smallness. A community had developed around Flaco, but that community was not immune from the usual human foibles. And if all was not well in Flacoland, it was about to get a lot worse.

AS A REPORTER, I am not above my own brand of smallness.

Before I visited Anke in her apartment, I had called her from North Carolina and suggested that after we conducted our interview we should tour the Upper West Side and track the last days of Flaco's life. It seemed like a nice way to wrap things up.

But I had made this invitation before I reached out to Alan Drogin, who lived above the alley where Flaco breathed his last breaths. When Alan invited me over on the same afternoon I was to interview Anke, I was in a bind. I emailed to ask if I could bring another Flaco follower along when I visited, but he wrote back: "Please do not bring anyone else. I've been tasked by our board to respect our privacy and limit visitors to respectable journalists and scientists. We decline requests for sentimental reasons and won't allow any photography."

And so, near the end of my interview with Anke, I tried to explain that she could not join me. I knew this was a reopening of the wound she had suffered during Flaco's last days, in fact an almost exact re-creation of that exclusion.

It is not to my credit that I made some excuse about having to meet Hadley after our interview, attempting to use my daughter as a human shield. Despite my attempts at obfuscation, Anke sniffed out what was happening. Another David was snubbing her.

"I WAS NOT REALLY a Flaco follower," Alan Drogin said while we sat in his living room. "I like looking for birds on my own. I have no big attraction to celebrity birds; I mean I saw Flaco I think maybe once in Central Park. That day a bunch of people were looking for him. There is a favorite tree he had across from the woodchip pile. And so, you know, I was just birding the North Woods and there were people there and I thought, *Well, what are they looking at?* And they were like, 'Oh that's the tree he's in' and I said 'Fine, okay,' and I looked up and I got to see him. I've seen eagle-owls before in South Africa. They're a little smaller there. They get bigger the further north you go, you know."

He took a sip of coffee and explained that after that he didn't feel the need to go search for Flaco again.

"Then sometime in February, I started getting woken up in the morning by a bunch of crows cawing. That's not too unusual. I've seen crows here, but usually they're single crows just flying back and forth along the riverside."

Alan's apartment is on West 89th, closer to the Hudson River than Central Park.

"It was odd. I heard these crows cawing about two or three mornings in a row. It sounded like a group of them. And I thought, *Oh, they must be going after something.* Though I wasn't a Flaco follower, I had heard some reports in bird circles that Flaco was further north, more like at 100th and 110th, around there, and that he was beginning to move a little bit further south. Meanwhile in our neighborhood there had been a peregrine falcon nest on 86th Street on the church on the corner of West End Avenue, just three blocks south. The birds had nested there the year before and had a successful brood, and there was talk that they were probably coming back. So I'm thinking, *Who knows, that might be what the crows are after*.

"So I decided to take a walk one day after I heard all the crow commotion just to see what was going on, maybe see if I could see what the crows were so excited about. And I actually found the crows two blocks away in a tree focusing their cawing on the top of a roof. There was a blue jay shouting, too, so I thought, *Okay there's something going on, something's around. Something suspicious*. And then when I was walking back home, I saw the peregrine falcon fly over one of the buildings. So of course I thought, *All right, that explains it. End of story. Mystery solved*. I come home, satisfied that I've figured it out, and just about dark, around four or five, all of a sudden I hear, *whooo*. And knowing what our local owls sound like I know that's not one of our local owls. So I'm hearing this and thinking, *Holy crap*, and I get out Merlin, and I start listening and thinking, *That's Flaco, he's hooting, he's around*."

Merlin is the app from Cornell Labs, the epicenter of ornithological study in this country, that helps you identify bird calls and songs.

"Sure enough my birder friends told me Flaco had been seen at 90th Street, which is literally a block away, on top of the buildings across Riverside Park. Okay, so he was in the neighborhood. After that I heard him maybe three or four more times. This would have been early February.

"It became a regular thing. I have good birding ears, so, you know, when we were sitting around watching TV at night I'd say to my wife, 'There he goes now, I hear him, you know, he's right out back, behind the apartment.' So I started looking for him. He was close, I knew that, up on top of the neighboring buildings. People were seeing him just around dusk, and he would be hooting up there. I told the super about it, and he joked and said, 'Yeah, great, maybe he'll take care of the rat problem.' And we left it at that. I spoke to Ralph the doorman about it, too, because he's a really friendly guy. And so, word started getting around the building that I'd been hearing him and that Flaco was around.

"That was it. And then I think within a few days, I ran into one of the people on our top floors with their kid in a stroller, and he said to me, 'Oh, we've got some pictures of the owl with our baby in the foreground as the bird is looking in the window.' It turned out Flaco was right across from our building.

"Flaco had started hanging out on the fire escape of the building next to us, so you could see him clearly if you were on the upper floors. He would sleep during the day on that fire escape. When I heard he was on the building next door, I went back down to the super, and I said, 'Can we go up to the roof?' And he said, 'Sure, you have access to the roof.' Which I never knew in all the years we've lived here—that we had access to the roof. So we go up to the roof, and sure enough, he's sitting right there, literally parallel with us, on the fire escape of the building next-door. My wife loves taking pictures, she's not a birder, but she loves it. She took some pictures, nice pictures, but he was sleeping so we left him alone."

Alan's experience was a kind of combination of all the different Flaco encounters I'd heard about. As with Nan, the bird had come to him, unbidden, but like David, or perhaps more like Bruce (who I didn't yet know was Alan's friend and his fellow member of the Linnaean Society) he was an elite birder who wanted to respect the bird's privacy. As for the photography aspect, his wife filled that role.

"That was the end of it, I thought. I'd heard him hooting on and off again after that, and then we stopped hearing him.

"And then one evening, the super rings me up and says, 'Flaco's in the alleyway, dead.'

"I said, 'Oh shit.' I ran downstairs, and it was a terrible sight to see this beautiful bird just spread-eagled out, face down. A huge bird, too. It was shocking really."

Alan stood up, put down his coffee cup, and said, "Let's go down and see the alley."

We took the elevator down to the basement. The super's office was right across from where the elevator opened, and Alan wanted to introduce me. But the super was off on some errand, so instead he introduced me to the super's daughter and granddaughter who were playing in his office. Growing up, the daughter had used the courtyard where Flaco died as a little playground, and lately the granddaughter had, too.

Courtyard, Alan had explained, was a fairly grand name for what was basically an alley between two tall apartment buildings. I now saw that was true. The area was all concrete, nothing green in sight, and it didn't at all match my vision of the relatively pastoral protected space I had imagined as the scene of Flaco's last conscious minutes on Earth. Two other alleys on the other sides of the building created a kind of concrete canyon feel. Staging, including scaffolding, ladders, and multiple platforms, covered the building next door, which was undergoing its five-year inspection.

Alan pointed to the spot where he had seen Flaco lying face down "spread-eagled." Then he narrated the fairly gruesome tale of the owl's demise. Apparently Flaco had fallen from the top of the building, ten floors up, or very near the top, and clipped the security camera on the way down. Later the super had to realign the camera from where it had moved after being hit.

"When I got down here, I noticed a little bit of twitch in the tail, like Flaco was still maybe alive. What I found out later was the super saw him very soon after he'd fallen, and he was still moving. Flaco had moved his head. But then the bird stopped moving, and the super figured he was dead.

"I called the Wild Bird Fund, and they didn't really register what I'd said at first. Whoever answered the phone didn't get that it was Flaco I was talking about. They essentially said, 'Oh we don't go collect birds, you have to get the bird and bring it to us.' I said, 'This is an owl. I am not going to pick up an owl with talons that could like, forget it . . . I'm not.'"

The person at the Wild Bird Fund suggested that Alan call the New York State Department of Environmental Protection, but it was evening by then, and he didn't know what department to ask for, even if he could get through. Miraculously, he somehow reached the right person, and that person gave him a list of people to call in the local area who would come and take animals. So he called the number and left a message, but then he called the Wild Bird

Fund back and this time it registered with whoever answered, 'Oh Jesus, you're talking about Flaco, aren't you?' Alan said, 'Yes. Thank you.' And they said, 'Okay, two of us will come right over.'

And so they came and took Flaco, and it was over, at least Alan thought it was.

But he was wrong.

"That was just the beginning. That was when the whole media circus began."

THE FIRST REPORTS WERE that Flaco had flown into the building.

First reports, it turns out, are hard to dislodge.

A building collision, a window collision. It made sense to most people. The media went with it. And went with it and went with it well into May and beyond. In fact as late as September, major outlets would still be perpetuating this narrative.

The result of the initial necropsy, performed by Bronx Zoo pathologists, was "death due to acute traumatic injury," with the main impact being to the body not the head, though there was some small bleeding behind the left eye. The report stated that Flaco had good muscling and adequate fat stores and basically weighed the same (4.1 lbs.) as he had the last time he was weighed at the zoo. In other words, he had been healthy right up until he wasn't.

The story that Flaco hit a building would persist, right down to a *Saturday Night Live* skit starring Flaco's widow. Buildings were the original culprit, and editorials were promptly written, with the New York State legislature even renaming a bird-collision law that had passed the year before "Flaco's Law." But while it's true that buildings kill *billions* of birds and that anything that might lessen this toll is admirable, David Barrett and others had their doubts. The fact that there were no massive injuries to Flaco's head argued against the collision theory.

If you saw the building itself, you would have been immediately skeptical. It wasn't a skyscraper that a migrating warbler might hit. It was a back alley, a concrete canyon, and had Flaco been flying any distance from one place to another, he would have had to dip down to even hit the building.

"A window collision, I doubt it," Alan said to me as I studied the relatively claustrophobic space where Flaco died. "This is a nocturnal animal. A nocturnal animal who is pretty good at flying around things. I'm all for turning lights off at night and changing windows. But this was in the back alley. Let's understand this was not a migrating bird."

David Barrett, who insisted on the same thing long before the final necropsy was in, had turned his considerable analytic powers toward piecing together Flaco's final hours.

"There were people all over the Upper West Side who would hear him hooting. So we know that he was a frequent hooter and hooted for the month before he died, often for many hours a night. But then suddenly beginning on the 19th, actually beginning probably the 18th, yes, beginning the evening of the 18th, no hoots. At least no hoots anyone heard. So that put us into thinking he was ill or injured, too ill to want to hoot.

"The initial necropsy showed there was traumatic damage found to Flaco, which was consistent with a fall onto concrete. The earliest reports in the papers and news outlets wrongly maintained he died from a collision with a building, but in fact that was never conclusively determined. It was just something that I think people quickly said because the injuries did appear to be consistent somewhat with what a bird would get crashing into a building.

"But not quite because he had no head injury. No head trauma. So that's an important clue that came out. Which made me think that a building collision was extremely unlikely. An owl wouldn't even have been flying fast enough to hurt himself by hitting a building because he was already inside the courtyard area. He wouldn't have been able to develop enough flying speed inside a building courtyard to do fatal injury to his body. So he would have been probably at most a few feet from the building when he was resting or about to take off. And he knew that building well. He'd been to that area many times before. He spent every night hopping around from building to building on the Upper West Side. He did it with skill and grace. He was not into making mistakes that would cost him his life. He could see perfectly well when healthy. And he knew how to get around buildings. That's how he spent his nights. He knew his territory.

"Millions of birds die from building collisions, but they're usually migratory birds. You can't call Flaco's death a building collision. The footage from the camera that night, the building security camera, showed Flaco falling pretty much straight down.

"My thinking that is consistent with all the factors is that Flaco was probably extremely ill or injured, and resting as he usually did on the windowsill of the building or on the railing, the fire escape railing of the building. And he probably continued doing that as long as he could. Birds can hang on a long time. They can lock on. But eventually Flaco became too weak to continue doing that. And that's when he fell. And he might have already been quite close to death when the fall began. And on the way down he could have hit a railing. He could have hit the railing underneath him. He could have hit a part of the building. [At this point David did not know he had hit the security camera itself.] Or he simply, and this is absolutely known, he simply hit the concrete ground, the floor of the courtyard.

"That in itself would cause traumatic impact. So, that's how he was found. He was found face down in that courtyard on concrete. And apparently still alive, but just barely. He showed wing movement."

IN THE WEEK AFTER Flaco's death things got crazy for Alan Drogin. The media circus had come to West 89th and had set up its tents at his apartment building. Reporters crowded the building. Flaco mourners placed flowers out front. The *New York Times* reported on Flaco's death, and several short essays were published in various outlets, usually with the word *freedom* prominently featured. At first the papers said the owl was a hero and he had died a hero's death. And of course he had been killed by *us*, our tall buildings, our poisons, and, to an increasingly vocal and adamant online contingent, our pestering. The message was clear. Flaco had died for our sins.

Because Alan was the building's resident birder and the one who, along with the super, had been with Flaco at the end, Alan's apartment board chose him to represent them. Which meant Alan spent the week playing defense.

"First you have all these reporters trying to jam microphones in your face, and then you have all these people with their own platforms. Taking advantage of getting media attention. So you have on one side the 'free animals from the zoo' people, using this as their chance to give speeches and then the Audubon Society saying, you know, we should turn our lights off and, you know, Flaco hit a window, and things like that."

The building residents wanted to protect their privacy, and Alan realized the potential trouble right away when one of the reporters from some television station started asking about their windows, whether or not they were the *right kind*. It wasn't a big jump from that to his saying, "No, our super didn't put rat poison out for anything." The board instructed building residents not to engage with any media but to funnel them toward Alan, who would try to respond properly. They decided not to show any photos of the demise to the press and didn't release the closed-circuit television film. When a respectable documentary filmmaker persisted, they showed him the footage, which looked like nothing, just something going fast by the camera.

To Alan it was all both ridiculous and morbid.

"They kept asking about Flaco hitting windows and I'm thinking, *No, Flaco didn't hit a window. Come on, cut that out*. But then again, I wanted to be diplomatic."

What Alan was realizing was that he was right in the middle of that modern phenomenon: the celebrity death. Flaco was Lady Di hounded by reporters (birders/crows) right to the end. He was John-John Kennedy, confused while navigating, flying off in a fog.

And that was just within the swirl of traditional media.

The online madness was worse, less restrained by the facts. There the mood swings of the media were magnified. In the wake of Flaco's death, there was much pointing of fingers, the usual online scoldings, critics adding a dark moral to the previously uplifting story. Things got nasty pretty quickly on X with a whole lot of angry *I-told-you-so*'s and *you-are-to-blame*'s flying around the ether. Any nuance went out the window. The "sentimental cat-loving types" who had made Flaco their hero were now somehow culpable, their softness getting in the way of hard science. Flaco had never been a wild bird, had never really felt free but had been scared and anxious the whole time he was outside his cage. So went the new narrative.

David Barrett was an obvious target, and for a while he, too, just played defense. Shell-shocked Flaco-loving posters grew a little more shy with the freedom metaphor. It would take some weeks for those who thought it was a good thing that Flaco had his year of freedom to cautiously emerge from their bunkers.

In the zoo's eyes at least the real culprit was clear. In their initial statement immediately after Flaco's death they wrote, "The vandal who damaged Flaco's exhibit jeopardized the safety of the bird and is ultimately responsible for his death. We are still hopeful that the NYPD, which is investigating the vandalism, will ultimately make an arrest."

This quickly became a popular sentiment online. Back when he had eaten his first brown rat, a petition had been started to keep Flaco free. Now a new petition was started: the vandals must be brought to justice. An online posse needed to be assembled. They had to get the bad guys. And of course there were bad guys. Someone was to blame. You got the feeling that if the original vandal or vandals, once seen by some as liberators, were found, they would have been strung up or at least put in the stocks in Grand Army Plaza so we could mock them and throw fruit.

THE FULL NECROPSY, CONDUCTED by Bronx Zoo pathologists and released a month later, revealed that Flaco, as well as suffering from pigeon herpesvirus from eating pigeons, was also chock-full of rodenticide, the poisons laid out for New York City rats. The Wildlife Conservation Society, which operates the Bronx Zoo as well as the Central Park Zoo, concluded, "Flaco's severe illness and death are ultimately attributed to a combination of factors—infectious disease, toxin exposures and traumatic injuries—that underscore the hazards faced by wild birds, especially in an urban setting."

"The autopsy confirms that Flaco was seriously ill with pigeon herpes, a disease that in these owls is 100 percent fatal within three to four days from onset," said David Barrett. "This explains why he ceased hooting in the days preceding his death. His fall from atop the building was a consequence of his strength and balance finally giving out."

The media, while still clinging to the building collision theory, now turned its attention to rats, despite the likely role that the pigeon virus had in causing Flaco's fall. There was a simple reason for this. Rat poison was a problem that you could fight against, and soon there was an attempt to pass a brand-new "Flaco's Law," focused on banning poisons and promoting something called "rat contraception." Despite the strange images the last phrase conjures up (rats putting on tiny rubbers), the media and many people took up the new cause with a vengeance.

I get it. I'm not sure what can be done about pigeon herpes, but rodenticide is another matter. It, like rats themselves, is a pervasive New York problem. Obviously it doesn't kill just celebrity birds, but many other city birds, including hawks and other raptors. And it is not just a city problem. Not long after Flaco's death, I reached out to my friend, author and ornithologist Scott Weidensaul, who has worked extensively with snowy owls through the research and conservation effort Project SNOWstorm. I was thinking back to that beautiful winter full of snow and snowys I had spent on Cape Cod and wondered if those radiant owls had been affected, too.

Stressing that this was just a preliminary analysis and that it is possible the numbers and percentages may change as Project SNOWstorm completes the statistical work, Scott wrote me back:

> Since 2013, Project SNOWstorm's wildlife veterinarians have assessed rodenticide levels in the livers of 196 snowy owls that died from various causes, and were salvaged by state or federal agencies and licensed rehabilitators. Thirty-five percent of those owls showed quantifiable levels of anticoagulants, meaning more

than trace levels. It's difficult to determine what a dangerous level of these very potent toxins is for a raptor, since it varies by sex, age, body size and physical condition, but research suggests that anything over just 0.03 ppm can lead to death. Of the owls we've tested, 44 birds, or 22 percent, were over that threshold, and almost all of those owls (93 percent) showed signs of internal bleeding, even those that had no other sign of trauma or injury.

What's even more worrisome, the percentage of snowy owls with rodenticide levels above the presumed mortality threshold has risen dramatically in the decade since we began this work—from near zero in 2013 to 56 percent in 2022. The extent to which this may reflect a growing exposure to rodenticides, or some other factor, is more than we can say at the moment. But it's clear that snowy owls are facing a significant toxicological risk when they come south from the Arctic for the winter.

IN PRIVATE, AWAY FROM the media circus, many Flaco followers experienced real grief. Anke wasn't the only one who responded to the news with tears. Thousands mourned.

Nan learned the news when she got a text from David Barrett.

"I knew when he texted me that morning it was bad. I was texting David back and forth and saying, 'Oh my god. No, no, no, no, no.' In his first text he told me Flaco was hurt but still alive. You know, I mean he was only alive for apparently a couple of seconds. And then David posted

Flaco has died, and I was shocked and stunned, but it wasn't until a couple days later that I started to cry and then I really cried."

It is a feeling of loss she wasn't able to shake.

"Not long after he died, Frank, the composer that I work with, sent me a new melody. He's doing an album of jazz standards, so he sent me this melody, and I wrote this lyric, and I sent it back to him. And he said, 'This isn't quite what I had in mind for a jazz standard.' And I read it over, and I realized that what I had written was a love song to Flaco. I mean it really was, it was a love song to an owl.

"There was this huge cave inside of me that—I remember thinking it's like a bullet hole. It was like an empty spot. And I remember writing to David, 'Well, if I have a bullet hole, you have a vast heart hole.'"

Unlike so many Flaco followers, Nan hadn't set foot in the park during Flaco's year, but she decided to attend the memorial being held at Flaco's favorite tree on the afternoon of March 3.

One of the reasons she went was she wanted to see David Barrett. While Flaco was alive, their relationship had been entirely by text, though they had met once the day after Flaco died. When she arrived at the memorial, she made her way through the crowd and right up to Flaco's favorite roosting tree. A woman on a microphone was talking about her love of Flaco. Nan asked around, but no one knew where David was.

"I wanted to see him in person. I figured he would be at the center of things, but he wasn't. And I kept texting him: 'Where are you?'"

He wrote back: "Walk toward the sun."

She looked around and sure enough the sun was starting to go down where it always did, over the buildings on the West Side. It was hard to see in the glare, but she walked in that direction and there he was, back a good distance from the crowd, under a tree.

"When I found David, I asked him, 'Why aren't you up there speaking?' Now I kind of understand why he didn't. He didn't want to be in the center of things. But on the other hand he gives all these interviews. He's a mystery."

Alan Drogin decided to attend the memorial as well. He still couldn't shake the upsetting image of Flaco dead on the ground and decided it might help to be with others who cared about Flaco. But what he found didn't inspire him.

"I would say only three or four of us actually had binoculars or were birders. And we watched as they're all having their moment of publicity. There is lots of media and cameras are clicking and they are talking about freeing animals from zoos, and it's the whole nine yards and everyone's leaving pictures and flowers and talking about how they love Flaco.

"I was actually standing pretty far back with a couple of birders I know, talking about the media circus, which was really upsetting to me. And then I swear while we were talking this guy with a boom mic moves close to us and without asking us tries to listen in on our conversation.

"That was too much. I didn't like all the stuff they were saying either. The causes. It was like going to your uncle's funeral and having people talk like it's an anti-war demonstration or something. So I decided, you know what, I'm going birding. That's what I need to do."

It turned out to be an inspired decision. One thing Alan had always loved about birding was the way you could get absorbed in it, no matter what else was going on in your life. He had worked as a computer programmer for years, and in his free time he loved sight-reading on the piano, and both programming and sight-reading shared something with bird-watching. You could get lost in them for hours. Fully absorbed. It was like a form of meditation.

"You're in the moment when you're going for birds. I'm a musician, so I love hearing the sounds, and you have to be very aware of looking around and seeing movements and things like that. It's the same with sight-reading piano—you cannot have anything else interfering, the only way you're going to be able to sight-read is if you concentrate on it entirely."

Unlike the memorial itself, birding seemed the right response to Flaco's death.

"It was what I needed to calm myself down. And it worked. I felt better. I came home. The next morning I got up to do some birding closer to my apartment in Riverside Park. Not long into my walk, I heard some screeching, and it was a red-tail. It was kind of dangling precariously on these twigs on the end of a tree with barely any leaves on it. *What's going on?* I thought. *What the hell is it doing?* Then I realized it was breaking twigs off to make a nest. And then sure enough, I hear another red-tail calling. So it's a pair, I think. They're making a nest. And so, I thought, *I want to end on a positive note here with the New York owl people.* I wanted to make my peace with them, to say to them, 'Look, here's a red-tail hawk, a pair of them.' They're going to make a nest. And they're not going to have an owl, a stranger, to compete with. And the owl's not going to have a red-tail to compete with. And we're back in balance here in the world. And in fact the red-tails wound up creating a nest at 102nd Street. And soon after their eggs hatched. So that's where I went with it."

As for Anke, she didn't make the funeral. She was up in Alaska photographing eagles. The pictures from her trip are stunning.

As it turned out she had her own sort of memorial. At four o'clock, the exact time the park memorial was starting, she was flying over Manhattan, back from Alaska. Which meant she had a bird's-eye view of the city and could look down at Flaco's territory, at the city he had called home for a year.

9

LEGACY

FLACO LIVES!

I WISH I'D seen him.

Three weeks after his death I visited the bird sanctuary in North Carolina where Flaco was born and spent a couple of hours staring at a pair of caged Eurasian eagle-owls. Beautiful birds that stared right back at me as I peered into their cage. One even flew close enough that I felt the wind of its wings, just as Sandra had. But it wasn't the same. I could never quite get there. There were bars between us. There was no electric current. Remember: all is context.

The enclosure wasn't as small as Flaco's, but it still seemed tiny for such large birds.

"In the zoo world birds don't get what they deserve," a West Coast zookeeper who specialized in birds told me.

As I watched the owls, I thought of an article I'd read titled "The Power of Patience" by Jennifer L. Roberts, an art history professor at Harvard. In her classes, Roberts requires of each of her students "an intensive research paper based on a single work of art of their own choosing." Students are required to visit a museum and spend three hours sitting and looking at their chosen work of art. Three hours! I decided that if the students could do it so could I, though I cut myself some slack and reduced three hours down to two. For two hours I watched the owls while they watched me. Among the things I noticed was the size difference between the pair, the female significantly bigger.

Another thing I learned, not long after my visit, was that Flaco and I shared a birthday: March 15. Beware the Ides of March. He had died three weeks shy of his fourteenth birthday.

By the time our shared birthday came around, I had embedded myself in the Central Park birding world and was spending every free minute of my days trying to tell Flaco's story. I was coming in at the end but talking to people about how it all began, a snake eating my snake tail.

Like everyone else, my personal Flaco story was colored by my preconceptions. My nearly lifelong quest had been to try and discover some form of wildness in a world that seemed to be doing everything it could to squeeze out the wild. Once, not long after moving to North Carolina, I witnessed a great cloud of northern gannets diving for fish out in the ocean. Gannets are beautiful white birds with six-and-a-half-foot wingspans and black-tipped wings that look like they've been dipped into a pot of ink. They stare down into the water from as far as a hundred feet up and then dive headfirst into the ocean where they tunnel underwater after fish. There were great clouds of them that day, and they shot into the water one after another—*thwuck, thwuck, thwuck*—like living lawn darts. It was thrilling and I, hungry for raw experience, took off my shirt and shoes and swam out close to where they were diving. From there, treading water, I nearly had a fish's point of view. I watched from below as they dove down to where I was, some of them landing only a few feet from me.

Why do such a crazy thing? To get beyond the thrice removed, to experience something visceral, to experience the thing itself if such a thing is possible. Of course this idea of "the thing itself" might just be another story we tell ourselves. A romantic story at that. An old hippie notion that gets lodged in some of our minds. But it is something I need to believe—something I cling to. The idea that occasionally and through grace and perhaps by putting ourselves in the right places (*read:* nature) we get to a place where our cluttered minds and cluttered places don't get in the way.

A more cynical take is that this was a stunt and I did it because I knew I was going to write about it. And if I am honest, I will admit the closest I get to Anke's excited obsessive state, and to the absorbed state Alan achieves while birding, is when I am writing a book. That is when it all comes together for me, everything connected.

I like the fact that while I am typing these mornings I get to listen to the barred owls. "Our" neighborhood owl hoots, and then another owl hoots back, often from across Hewlett's Creek, the tidal waterway we live on. Hewlett's Creek turns out to be Dawson's Creek, where the TV show was filmed, so even though I am in North Carolina in a way

I still live on Cape Cod, albeit a fictional version. In the show the Dawson character lived on the wealthy side of the creek, and his love interest, the character played by Katie Holmes, lived on the other. The two barred owls seem to be replaying a similar drama, hooting back and forth across the creek.

As well as watching owls and writing owls, I have been reading owls, books piling up on my desk. These include David Sibley's *What It's Like to Be a Bird* and Carl Safina's *Alfie and Me*, about the screech owl he raised, as well as a few books by Jennifer Ackerman, including *What an Owl Knows* and *The Genius of Birds.* For boning up on my local barred owls, I turn to Scott Weidensaul, who compiled the *Peterson Reference Guide to Owls of North America and the Caribbean*, and when I am feeling a tad more mystical, I dip into *Shamanism: Archaic Techniques of Ecstasy* by the renowned religious scholar Mircea Eliade. I learn that for shamans a "bird costume is indispensable to flight to the other world" but decide not to don feathers and stick with my sweats and t-shirt, perhaps not quite ready for that sort of prolonged flight. The book I feel most kinship with is *The Wise Hours* by Miriam Darlington, who interweaves her study of owls with her life and gives the whole a feeling of personal quest.

One thing that Darlington's book makes clear is that we have long foisted our stories on owls. Often those stories have to do with our fears regarding the birds, due in part to the haunting nature of their hoots, the hours they keep, and those giant eyes that seem to look right into you.

Darlington writes: "In Egyptian, Celtic, and Hindu cultures, the owl's symbolism was involved with guardianship of the underworld. The owl was revered as the winged keeper of souls after death. In Malaysia and Indonesia it is *burung hantu*, the 'ghost bird.' The ancient Greeks associated it with wisdom and courage; the Romans with foreboding and fear. Wise or evil, the owl was a porous receptacle for all of our chosen meanings."

She continues: "In Japan, on the other hand, the word for owl is *fukuro*, which means 'good fortune,' and so the owl is lucky. In Aboriginal Australia, Eerin the grey owl is a protector, sleeping by day and flying by night to keep watch and to warn if danger approaches. In South Australia the Nyungar tribe protects a standing owl stone, Boyay Gogomat, a creator, healer, and destroyer. The Wardaman tribe in Northern Australia believes that at a unique rocky outcrop that overlooks the outback dwells Gordol, the owl who created the world."

And so on.

The meanings vary. The one commonality is that owls have always fascinated human beings.

"I understand the attraction of owls," Alan Drogin told me during my visit to his apartment. "When I go birding and take a group into Central Park, if we see an owl, like a barred owl or something, I will lose half the people on the walk. They will just stay there and stare. It's like, you know, this attraction they've got. It's partly the anthropomorphizing human quality—the eyes in front. And this happens all the time. Half the group gone! I'll just keep leading us on, and I'll circle back later and they're all still there sticking around looking up at the owl."

"People have an extreme fascination with owls," concurs Dr. Andrew Farnsworth. "And that actually transcends whether you know nothing about birds, whether you are an expert, whether you are a scientist or a photographer. And owls have this incredible ability to bring out absolute best and worst traits in birders and nonbirders. There are countless people who experience an owl for the first time, whether they hear it, usually when they see it, and say, 'Oh my God, this thing is incredible,' and it's the gateway to engagement."

Farnsworth is a visiting scientist at Cornell who studies bird migration. He is widely respected in the birding world, and so when I called him, I expected him to be critical of the amateurs who followed Flaco. That was not the case. Instead he seemed to delight in the fact that thousands, maybe millions of people had discovered the pleasures of watching an animal.

"I think it is a great thing that so many people became engaged. The world has changed. What used to be exclusive knowledge is now open to everyone."

Of course he understands the dangers. He spoke of the potential "tragedy of the commons," the way that when some subset of a larger group overuses a resource it is no longer available to the whole. In this case, the commons are Central Park, the resource is Flaco, and the threat is those who might crowd or disturb a rare bird. But he also mentioned something I hadn't given much thought to.

"There are always bad actors," he said. "But there is also a kind of self-policing in a group, particularly as people learn more."

While there were exceptions, in the case of Flaco a certain etiquette was established—the old pros warning the newcomers about getting too close or disrupting the owl.

Curious, I pushed him toward being a little more critical of the world's Birding Bobs. He admitted that the use of regular playback calls could disturb the daily patterns of birds, but he also said that the science wasn't clear on how much of a disturbance it really was.

In particular, Dr. Farnsworth stood by his friend David Barrett. Though he was aware of the dangers, Farnsworth clearly favored the diverse multitudes learning something new over an elite group holding tight to secret knowledge. Not that this would instantly transform casual observers into ardent environmentalists or even into birders, but it might open some eyes to the importance of lives other than the human.

NEAR THE END OF May two bluebirds, a male and a female, came to the large plate-glass window that faces our backyard. The male attacked the window, likely seeing his own reflection. Nina put a blanket over the window, but that didn't help. He found another smaller window, and when we blocked that, another. The bluebird of unhappiness, we dubbed him.

Similar displays of territoriality, with similar results, were on display on X all spring. As a newcomer, I had learned that it is a world that could create communities, but it is also a world where magnanimity and simple good manners are often pushed aside. In the wake of Flaco's death, someone had to be blamed, and posters pecked at each other as if it would heal their own wounds. At one point, as I began to write about Flaco, I also began to wade into this world. In one of my first posts I suggested there were two sides, at the very least, to the story of Flaco's last days. To make my point I quoted the famous F. Scott Fitzgerald line, "The test of a first-rate intelligence is the ability to hold two opposed ideas in the mind at the same time, and still retain the ability to function." Nobody on either side liked the quote very much. They had their stories, and they were sticking to them. I realized that if that is really the test, most of us are failing.

For a lot of people, Flaco's death and the grim necropsy cast a pall over the whole Flaco fable. And perhaps it does. But a fable can mean more than one thing.

Just as Flaco himself was a bird of a hundred expressions —one moment pensive, the other curious, the other seemingly outraged—and a hundred postures—scratching his face with those deadly talons, stretching and lifting his wings, that versatile neck of his always bobbing and lifting and doing figure eights and swiveling and pivoting—he also inspired hundreds of stories.

Flaco even managed to get caught up in the culture wars. On February 25, just two days after his death, Nicole Gelinas, writing in the *New York Post*, pointed to Flaco as "another example of NYC progressivism gone awry." She wrote that "well-meaning, naïve New Yorkers have spun the owl's story as a feel-good tale," but it was really "a tale of vicious vandals . . . maliciously abusing a defenseless animal, abuse that resulted in that animal's cruel killing." Gelinas had great disdain for the soft liberals who saw the bird as a symbol of freedom and "a feel good meme," and who took pictures of him with their "always-at-hand ready-to prey 'mobile devices.'" (You know she was tempted to use the word *newfangled.*) In this version Flaco's year was full of fear, distress, and disorientation. And of course it was a tale of crime in New York City and the need to bring the vandals to justice. "Let's hope that District Attorney Alvin Bragg ensures that he, she or they do real jail time," Gelinas concluded. "If the animal cruelty law doesn't carry a harsh-enough penalty, felony burglary and felony larceny—when actually prosecuted—certainly do."

Another subgenre of Flaco lore is what you might call Flaco-as-Rachel-Carson. Throughout the year this eco-thread ran like an underground stream below the main narrative, but with his death it broke above ground and grew wider and stronger. This was Flaco as the canary in the coal mine, warning us about the threats that all birds face. The first of these threats to be spotlighted in the days after Flaco's death was of course the story of building and window collisions. While that was not what killed Flaco, the misperception did bring attention to an enormous problem.

Consider that three billion birds—three *billion*!—almost a third of all the adult birds in North America, have gone missing on our watch over the last few decades. The threats that have led to this great diminishment are varied, but building collisions kill around half a billion birds a year in the U.S. This happens because during the fall and spring migrations, when birds often travel at night, lights disorient them, and the glass of windows and the sides of buildings reflect back a world hard to distinguish from the actual one they are flying through. While most environmental problems seem to have no answers, making us want to turn away or curl up into mental fetal positions, the answers to this problem are actually pretty simple. The National Audubon Society suggests three simple actions: dim your lights during spring and fall migrations, make your glass windows "bird friendly" by breaking up the glass with strings or dots (or simply closing your blinds), and help push for a lights-out movement during migration, a movement that has already had success in several big cities. So, while Flaco didn't hit a window, he might have ended up preventing other birds from doing so. At the very least, New York City, and to some extent the world, had never been as focused on the problem as it was during the weeks after Flaco's death.

Then there was rat poison. Once again this is probably not what finally did Flaco in, but he did have four different strains of the poisons in his system when he died. Having gotten up close and personal with several New York City rats during my visits to my daughter, I understand the urge to get rid of them. But, as Dr. Afshan Khan said during our Zoom meetings, it is inhumane and, just as with Rachel Carson's fable of DDT—the chemical that worked its way up from the insects it was meant to kill through the marshes to the birds—rat poison doesn't just end with the rats. It gets into the water, into pets (the most recent example being a seven-month-old Rottweiler puppy who died after nibbling on rat poison during a routine morning walk with her owner through their Manhattan neighborhood), into ourselves, and, it turns out, into eagle-owls and other birds.

In the months after Flaco's death, in good part inspired by his death, parts of New York City were still considering a bill banning rat poison. The new law would replace the poison with contraceptives that are stuffed inside salty fatty pellets left out in rat-infested areas. According to the *New York Times*, the contraceptive works by "targeting ovarian function in female rats and disrupting sperm cell production in males."

So now if you google "Flaco's law" you will find your way not just to one but two laws. First, the previously passed "Bird Safe Building Act," which was renamed the FLACO ("Feathered Lives Also Count") Act, mandates a lights-out policy in all government buildings. And second, the rat poison ban in two select neighborhoods. Pretty impressive lobbying for a single bird.

As a footnote, I should add that all the articles reporting on these bills never failed to mention that Flaco had died because he flew into a building. A secondary cause was rat poison.

Lucky pigeons. They got off easy.

I WASN'T ALONE IN trying to tell Flaco's story; narratives filled the air. Sandra told me at least three documentaries were being made. David Lei and Jacqueline Emery, now very much a couple, were writing a book together. Anke also dreamed of creating a coffee table book made up of her beautiful photos. Meanwhile Nan was hard at work on a children's book about Flaco's escape.

Like Flaco himself, the stories were ever changeable. In late May I went out to dinner with Nan and her husband, John, a former entertainment lawyer turned producer. In the course of a delightful meal, I learned something new about the day of Nan's Flaco visitation. What I learned was the simple fact that Nan was not alone when Flaco came to her. It's true she had been by herself in the kitchen when the owl first arrived, but soon after she ran to get John, and he had been there almost the whole time Flaco was at their window. I'm not sure why that seemed so shocking to me. Maybe because of the way Nan first told me the story, I had imagined it as a very personal, one-on-one encounter between woman and bird, something intimate and individual. But of course John had his story of that day, too, and though he might not have felt the lasting bond with Flaco that Nan did, he had been delighted and lifted by the experience.

It turned out everyone had their own Flaco.

Consider the old-fashioned morality tale Bruce Yolton was telling over on his *Urban Hawks* blog. On March 25, a month after Flaco's death, Bruce wrote:

> The postmortem lab results had come back for Flaco. You can read the details on the Wildlife Conservation Society (WCS) website. The testing revealed two underlying causes that contributed to his death, a herpesvirus most likely transmitted from his prey of Rock Pigeons, and exposure to four different rodenticides due to his consumption of Brown Rats. His cause of death, which was initially linked to a collision or fall, was precipitated by these two underlying conditions. As I said a few weeks ago, what really killed Flaco was science denial. While the vandal who released him is ultimately responsible for Flaco's death, there were many organizations that failed Flaco. The zoo failed to pro-

> tect and recover Flaco. The Wildlife Conservation Society (WSC) has never addressed the poor security at the Central Park Zoo, the public relations vacuum that led to the disruption of the rescue attempts, and the abandonment of the rescue attempts after only two weeks which the WCS justified because of Flaco's ability to feed himself, ignoring the risks he faced in Manhattan and the risks he posed to native wildlife.

He had the right to a little I-told-you-so since in many ways Bruce's warnings throughout the Flaco year were prophetic. "It sucks being right," another ornithologist told Alan after Flaco's death, acknowledging the sadness of it all but also that the writing had always been on the wall.

Unsurprisingly Bruce's story had a villain and the villain's name was David Barrett:

> Social Media influencers, such as David Barrett, who actively interfered with the rescue efforts to recover Flaco also share responsibility for Flaco's death. As does both the print and television media, who rather than calling for Flaco's recapture, glorified his existence outside of the zoo as some wonderful expression of freedom and an immigrant story of someone arriving in the Big Apple and making a new life.

While this sort of criticism must have privately pained David Barrett, he outwardly shrugged it off. It was the price of social media fame. Like those "popular singers" he mentioned but couldn't name. "Much of the world loves but a minority hates. It's just how things work out." And certainly he was not wrong to think that jealousy had something to do with the venom of his critics.

Through it all David didn't waver and stuck to the story he told me the very first time we spoke: "We created a Flaco community, people who really wanted their fix of Flaco every day, who wanted to know he was well. People worried about Flaco. I worried about Flaco every day. So, we always wanted to reassure people with a photo of Flaco, saying, 'Yes, he's fine. He's in the park,' or wherever. For some people that meant going to the park to see him for themselves. But you could also really get a daily look into the life of Flaco without ever leaving your mobile or your computer.

"One of the things that made Flaco so popular was that he was an individual. So many birds are anonymous. They're beautiful, but they look just the same as a thousand others of their species. Flaco was distinct. He was the only Eurasian eagle-owl in the wild in all of North America. And he was a bird that people could come out and see as an individual every day and follow.

"That was one of the things that made him a celebrity bird. There are other beautiful birds. We photograph beautiful birds all the time. It's just that people don't connect with a specific one because they're part of thousands that look all the same. Flaco and other owls like him are distinctive that way because we only have one in the park usually.

"By having people feel a personal connection to Flaco, people were more likely to feel a personal connection with nature. That's one of the things my account tries to do. I try to get people to get interested in particular birds, 'celebrity birds' we call them, not only because that makes for viewers and a great viewership, but also because it gets people to like all the other stuff I do, too, and they become nature aficionados, appreciators of nature, and are more likely to think favorably of conservation of nature."

I have some stylistic differences with David, and I'm well aware that ego and a craving for a certain type of fame is mixed in with his Flaco obsession. He's no Anke. He won't be swimming out to watch gannets dive or slicing his arm open to slip an osprey bone under his skin to enhance hunting skills anytime soon. On the other hand, there is a reason that *Sherlock* was his favorite TV show and he was expert at sifting through the clues and tracking down Flaco, and in the course of the year he made himself an expert in Flaco in particular and eagle-owls in general. He threw his whole self into his Flaco pursuit and in doing so brought people, and not just committed birders, into Flaco's world. He did that for people near Central Park and for people in Tokyo, Brazil, and Slovenia.

Whatever else you want to say about David Barrett, he helped thousands, maybe tens of thousands of human beings find Flaco. One of the things Flaco's huge orange eyes could always do was draw people out of themselves. And while they were watching Flaco these people were focused not on their reflection in the mirror or on the latest Trump news or even on their own to-do lists but on the fate of an individual bird. (I would say they were looking at a bird instead of their phones, but the truth is that they were often enough looking back and forth between their phones and the bird or taking pictures with their phones or even, for those far away, looking at the bird on their phones. But the focus was generally avian.)

And that was one of the most profound aspects of the Flaco story. It showed us that human beings can care about something other than human beings. We are all so wrapped up in our dramas, our personal dramas, yes, but also our political and societal dramas, that we forget to look beyond ourselves. We are too often sunk in a species-wide narcissism called anthropocentrism.

The Flaco story had self-promotion and silliness and was full of media overkill and blame, but it was also the story of humans, a whole lot of them, looking out toward an animal. This I find encouraging. Not world-saving but encouraging. Simply looking outward.

ZOOS HAVE ATTEMPTED, WITH varying degrees of success, to evolve with the times. There are even zoos that employ some of the same language Jack Turner used in *The Abstract Wild*, proposing that we need to rewild our zoos, making animals' environments much closer to those they evolved in. The metaphor of the ark has long been a prevailing one for zoos, and there are many historic examples of their preservation of endangered species. Writing in the anthology *The Ark and Beyond*, Henry W. Greene, an emeritus professor of ecology and evolutionary biology at Cornell, considers the attempts of zoos to become wilder. He writes that places are "wilder to the extent that they entail locale-appropriate ecological and evolutionary processes, including predation, competition, selection, disturbance, and nutrient cycling."

Therefore "maximum wild ecosystems must contain organisms that enable those processes, including apex predators, megaherbivores, and scavengers."

Life in the wild is complex and dangerous, and its challenges enlarge the animals, sometimes literally, in the case of wild lions having larger braincases than zoo-raised lions. Greene asks, "Wouldn't rewilded zoos better serve conservation, research, and education through more realistic and interesting portrayals of ecological and evolutionary processes?" To say nothing of the experience of the animals themselves.

During the Flaco saga the Central Park Zoo and zoos in general took a beating, and no one was talking much about the more noble work zoos do, that of species preservation and conservation research. Competing with these goals is the simple fact that zoos are businesses that have to survive, and so one of their primary goals is to entertain.

Another primary rationale for zoos is that they are not just entertainers but educators. In this they sound a whole lot like David Barrett when he says he hopes Flaco-watchers will become "appreciators of nature" and of conservation efforts.

One thing almost everyone I spoke to, including David Barrett and Bruce Yolton, agreed on was that the Wildlife Conservation Society, who run the zoo, lost control of the narrative. Perhaps this was out of defensiveness once the freedom narrative took hold and people started noticing how small Flaco's enclosure had been. After the WCS decided not to try and capture Flaco, they also decided to say very little, and they clearly made a mistake by not communicating, that is, by not staying ahead of or at least keeping up with the story.

Dr. Andrew Farnsworth also agreed that the WCS and the zoo had "really missed an opportunity."

"I like WCS, and a lot of amazing things come through that organization. But by not talking they didn't do themselves any favors. Instead they could have said, 'What story can we tell?' They could have led weekly walks to see Flaco, right?"

An irony here is that another primary goal for zoos is to educate by having people observe animals. The WCS really did seem to miss a great opportunity in Flaco, whose location was known for most of the twelve months he was free. In a way Flaco-watchers were experiencing a movable zoo. One way to look at David Barrett's Manhattan Bird Alert was that it was filling a void that the zoo left.

Another idea that a lot of people ended up agreeing on was that of sanctuary. The more I heard people mention this, the more it made sense. Capturing Flaco and returning him to a place where he could live without the threats of rat poison, cars, buildings, and crowds. But where would that place be? One of the Zoomers had suggested the Adirondacks, but there he would have threatened native wildlife and competed with native raptors. Was a sedated flight back to Europe really a possibility? And where would he be released?

And what was Flaco's natural habitat? David Lei had traveled to Spain soon after Flaco's death and photographed plenty of eagle-owls, but most of these were not in sanctuaries, or even in any sort of wilderness but were urban dwellers like Flaco. The adopted habitat of these birds seemed to be a lot like Flaco's during his year in the city.

"I do believe in sanctuary," Alan Drogin had said. "Whether you are for zoos or not, if you're going to have these captive animals, especially raised by us, you don't just let them out in the wild for nothing. You have to find some place for them, like the elephants that now have a place in Tennessee. It's true that not all sanctuaries work. But you can't let them just go on their own. You have to give them a chance to survive."

"It's a hard choice," Anke told me. "I don't think that either one was for its best. It wasn't best for him to be back in that horrible cage. But it certainly has its dangers to be in Manhattan as well. I actually hoped that Flaco, if he grew weak, would be captured and helped. You could capture him, you could rehabilitate him, and then put him somewhere in a bird sanctuary outside of the city, where he could live the rest of his life, in relative freedom. That was my hope. That didn't work out."

It is worth noting that the support for moving Flaco to a sanctuary, while popular after his death, might well have been greeted with outrage by the Free Flaco contingent when he was still alive.

WHILE EVERYONE ELSE IS telling stories, Flaco will never get to weigh in as to whether his year in the urban wild was worth it or not—though some would argue he did weigh in, not with words but with his actions. As my source at the zoo told me, his enclosure was kept open as an invitation for the year he was gone. This was not an invitation Flaco took.

It is not uncommon for animals to return to their former homes, but Flaco, other than a few days early on loitering near the zoo, did not. He refused the invitation and steered clear of the space where he had spent all those years.

Was his year filled with the exhilaration of freedom? Or was it filled with fear and anxiety? Most likely it was the usual mix we all endure, the sloppy quotidian combination with which all animals, domestic or wild, human or otherwise, make our way through our days. But of this, like so much else, we can't be sure.

Since we can't know, we imagine, and, in a move that will perhaps lump me with "the cat-loving sentimentalists," I choose to remember Flaco in flight. Not plummeting downward from his perch, propelled only by his own weight before hitting a security camera, but lifting off of a branch on his favorite oak, pushing off into the sky, the startling white of his underwings and his always-surprising and impressive wingspan on full display. Before he lifts off, he makes that little rocking motion and pushes, the kind of motion that humans know from diving into a pool.

"I was lucky enough to see him figure things out," David Lei said after Flaco died. "Just to see him soar from one twenty-story building to the next really spoke to me about how far he'd come since he was an owl on the sidewalk of Fifth Avenue."

That is the moment we would like to remember, isn't it? The moment of growth, of becoming more after all those caged years. This is the moment that all the early Flaco followers—Bruce, Anke, the Davids, all the rest—whatever their later divisions, witnessed. A moment when, after weeks of hunting rats on the ground, studying them and hopping along after them in almost comic fashion, a new moment came, not comic anymore but dramatic, the moment when Flaco pushed off a branch and silently swooped down, dropping, casting a shadow over his frightened prey, and then cleanly plucked up that prey in his talons, clutching it while, in a manner that would have been so foreign to him just weeks before, he gracefully rose and, flapping and curving and lifting up, alighted on the branch of a different tree.

In the end all we have are our moments, and if we are lucky, our days. That would have been a good moment, a good day. Something he could not have imagined during his life of dull confinement. He had broken beyond what had been. It was a new world. For a moment he was living uncaged.

EPILOGUE
BEYOND FLACO

FOR SIX MONTHS now I have been an observer, and not just an observer but an observer of those who observe, doubly removed. It is time to cut out the middleman. I am hungry for some wildness of my own.

It is time for me to see an eagle-owl in the wild. I remain a writer, of course, and my obsession has been as much or more with telling Flaco's story as with Flaco himself. But as spring has become summer, I have begun to realize I am jealous.

Jealous of Anke's obsession with Flaco, jealous of Nan's visitation, jealous of all that the Davids have learned about eagle-owls, jealous of everyone who got to see the bird in the park and in the city.

It is the very last day of July when I attempt to rectify this feeling, trying to fill the void. I do so in an extreme way, boarding a plane from JFK airport to Helsinki. Back in May, an environmental writer from Finland, Juha Kauppinen, visited me in North Carolina and spent an evening on the back deck of my writing shack, drinking beers and swapping stories. Later that month, as I started to read more deeply about owls, I began to see that many of the owl experts, mostly bearded older men who could have stepped right out of the pages of *Lord of the Rings*, reside in Finland. And that eagle-owls reside there, too.

It turns out that one of these bearded men, Heikki Willamo, is both a friend of Juha's and one of the country's best nature photographers, and that he spent several years back in the early 2000s climbing up to a cliff he called Eagle-Owl Hill, where he photographed a pair of eagle-owls and their young. Before long I have concocted a plan to fly to Helsinki and climb that hill with Heikki.

It is August 2, almost six months after Flaco's death, when Heikki, Juha, and I begin the climb. An hour earlier we had arrived at Heikki's nearby home, a beautiful converted schoolhouse with spacious rooms and outbuildings with moss roofs an hour northwest of Helsinki. When white-bearded Heikki said, "I will be your shaman," he wore a wry smile. It was a joke, at least partly. But now, walking down a dirt road with farmland on one side and woods that guard a granite cliff on the other, we are nervous. What if I have come all this way and don't see a bird? There is no guarantee that I will. Heikki and Juha feel the pressure, too.

A dozen common cranes, beautiful long-necked white birds with black wingtips, black half-patches around their eyes, and red markings on their heads, graze on seed out on the fields, looking anything but common to me. Then, after I comment on how unusual it is for someone from my country to see stands of aspens and birch trees growing together in large numbers, like Vermont and Colorado squashed together, I look up and notice a distinct dark shape rowing with deep flaps across the sky from cliff to woods.

"An immature," Heikki says, "one of the young," and just like that, the pressure is off. I will not go 0 for Finland. I have seen my first eagle-owl in the wild.

We laugh, relieved. Heikki now knows he will not fail in his shamanistic duties. Soon the road turns toward the cliff, and we take a path that winds upward through a hobbit world of reindeer lichen, moss, and stone. *Mythical Journey* is the title of one of Heikki's books, and already this trip is taking on some of that vibe, especially when we climb out of the woods and up into the open space atop the cliff.

Heikki joked earlier that this was what qualified for a mountain in Finland; "Our Rockies," he said. But while it is a fairly easy climb, the view is as satisfying in its fashion as those I have seen from the Continental Divide. Our patio is a ledge of granite and moss with a drop-off of maybe a hundred feet, where we stare out at a mix of farmland and forest. Heikki breaks out three Sandels, a brand of Finnish beer, and we toast our ascent.

We listen to the tinkling music of aspen leaves, and I realize that in the fall this whole hillside must be lit up gold. We drink our beer and stare out at the fields, soon spotting a moose—"elk" they call it—and a roe deer, an animal I have only heard of because the books I have recently been immersed in tell me that eagle-owls have been known to hunt them. In the woods below three fledgling eagle-owls screech, begging for dinner. It is a sharp, grating, insistent sound, a sound that fits its goal of rousting a parent to hunt. Heikki explains that during nesting the female tends to the young while the male hunts, but now most of the work, the hunting and feeding, falls to the male.

Juha and I continue to scan the fields with our binoculars, while Heikki heads back into the woods behind us, to urinate, I assume. But whatever he goes in for, it is what he comes back with that counts.

He returns and smiles and points behind him.

"Look," he says simply, and we look. What we see, not thirty feet away, is a large eagle-owl in a fir tree staring down at us.

The light is fading, but its eyes glow orange. Its puffed chest is slashed with dark marks that look cinnamon in the light. I feel a mix of disbelief and wild excitement. It is the very thing I have been thinking about studying, reporting on, but now it is real. It stares directly at me. If I were half my size, I would be nervous.

It is the beginning of a magical two hours. Crepuscular hours, though dusk this far north is close to 10 p.m. We have all made jokes earlier about words like *magical* and *shaman*, but this is exactly what I have come for, this experience as much as the sight of the bird, and there is magic to it. I have experienced times like this before in my life, times when the sights keep coming, the light keeps getting better, when you are as full as you can be and then suddenly fuller, like a gambler on a roll.

The owl drops off the fir tree, flaps, showing us the white undersides of its wings, and glides over to the top of one of two tall spruce trees that Heikki calls "the twin towers." Two decades ago Heikki spent every day here, photographing these birds and getting to know them personally. The bird we are watching, he tells me, is a male. He is twenty-two years old. Not close to the record for an eagle-owl, but long-lived for a bird.

Darkness is moving in now, coming in like the tide in almost imperceptible shifts. The young continue to beg. We hear a barking below that Juha first thinks might be a lynx, but then he apologizes, no, it is likely just a roe deer. This seems to be confirmed when we see a deer freeze out in the field and we hear more barking, which we suspect is a warning from another deer letting the first one know something else is nearby. We are distracted from that drama by another, as the owl swoops down off the spruce tree and flies to a gnarled pine not fifteen feet from us.

If I have not made this clear already let me say it once more: to see an owl is different from seeing other birds. Part of that is the way they truly stare right into your eyes. That is what this one is doing now. I look at him without my binoculars, and then, though he is close, I stare through the binoculars at that stark and exaggerated version of his face, almost human, almost feline, those burning eyes, the facial disc that aids vision as well as hearing, the shoulders huffing up and then relaxing, the adjustment of the head as it bobs and weaves.

Having paid us a visit, he flies back to the tallest spruce, and as the light fades we consider the possibility that we might have had enough and should leave the owl in peace. He has work to do after all. Down below we can still hear the young birds screeching for food. It is time to go and let their father fulfill their demands.

Then, just as we are getting ready to leave, he drops off the spruce. He doesn't flap at first, dropping like a stone, and I find myself thinking that that is exactly how I would drop off that tree. But then with a single flap he is gliding in front of us and then down into the valley. He is after something this time, you can tell. Juha lifts his camera. Heikki and I, he with his bare eyes and me with my binoculars, try to follow the dive. As the owl gets close, we see something. "A rabbit," one of us calls out. But it is not a rabbit.

What the eagle-owl is diving toward is much larger than a rabbit. It is a lynx.

He is not likely hunting the lynx but warning it. *Get out of my territory. I have work to do.*

We do not see all this of course. Just a blur. But a few seconds later Juha, who by some miracle has captured the moment, is showing us the viewfinder of his camera. He is also yelling with excitement. There is the owl, wings spread wide, and there is the lynx, black tail and white paws, running off.

This evidence adds to our growing elation. The owl perches in a tree above the field where the lynx was, far below us. Juha tells me that there is another name for lynx in Finnish, which he translates as "shadow that does tricks." The shadows themselves have been swallowed by full darkness. It is also suddenly quieter, and we realize that the beseeching of the young owls stopped as soon as their father dropped off his perch and swooped into the valley. I drink another beer and pace back and forth on our granite stage. We are not ready to leave.

SHAMAN NUMBER TWO IS wilder.

On my first afternoon in Finland Juha and I walked through the Kotinen Primeval Area, a deep mossy woods near his house that he called "my cathedral." Marta, the family's black Lab, came with us, and I marveled at the way she tore through the woods, leaping over logs and ripping through thickets and thorny briars, somehow never getting injured.

Pertti Saurolo, though eighty-five years old, has a style similar to Marta's. Finland's premier osprey expert charges through the woods in his knee-length raincoat and rubber boots. He has long white hair and a white beard in the manner of Santa Claus, but this is no roly-poly elf. Rather he is angular and wired, jumping over logs and cutting through brush like an athletic Gandalf. I, wearing the same sort of getup as he is, coat and boots, try to keep up. Rain is pouring. "It's deep there," Pertti yells back as he leaps over a small gully onto a landing pad of peat. These mossy, dripping woods remind me of woods where I once spent a week in a cabin in Oregon.

An hour ago Pertti dumped some maps in my lap and told me to navigate as he drove his old car through a maze of forest roads. Now he continues to charge through thickets over the peatlands and bogs. My boot is almost sucked off when I land in one of the boggy gulleys. I am aware of the comic aspects of our mission, but I am also tired and uncertain and, thanks to jet lag, a little anxious. I stuff those feelings down. It seems we are truly lost for a while, though Pertti has a compass and GPS. "Five hundred meters that way," he yells back to me while pointing ahead. He says there's a path nearby, and I don't really believe him. But then, miraculously, a path appears.

After about another hundred yards, or rather about another hundred meters, he stops.

"You see it?" he asks.

I don't at first but then move a couple steps closer, and there it is, an osprey nest, maybe eighteen feet high in a tree. Though the bulky nest is not on a platform like so many of our nests in the United States, there are some wooden supports at the top holding it up.

"A good storm could take it down," Pertti says.

Our job today is to survey the nest and determine if it is occupied. Earlier in the year an assistant of Pertti's reported it was not, and at first that seems to be the case. But then Pertti starts to whistle.

I know what an osprey call sounds like and can do a pretty fair imitation myself. This, however, is a whole different level. This is musical genius, a series of rising and cascading whistles that sound as much like an osprey as an osprey does. Which the birds obviously think, too, because soon one osprey, a juvenile, and then another, the adult female, are circling above us. We stare up through the trees. Another osprey calls from farther off. In any other nature situation with an expert, I would admit to being a bumbler and would defer. But with ospreys I can actually be helpful, aiding with the identification. Pertti is pleased with what we have discovered.

The rain lets up. Before we move on to survey the next nest, I point up into the tree at the nest. I know Pertti has climbed most of these trees to band the nestlings and ask if he climbed this one.

"Yes," he says. "This is an easy tree. An extremely easy tree. But it is a widow-maker. Its top can break at any time. I climbed up last year to ring the nestlings when I was eighty-four. But this year I got sensible enough not to."

"You're just going to send me up there," I say.

"No," he says emphatically. "No one will be climbing the tree. If anyone will be climbing the tree, it's me."

The reason he will not climb it today is not really because he has gotten sensible. It is because of a recent operation on his knee. He says he hopes to be climbing again next year.

"I would like to be climbing a week before I die," he adds.

And then he is off, charging through the wet woods back to the car so we can drive to a spot deeper in with access to the next nest.

WHY GO ON ABOUT ospreys? What do they have to do with eagle-owls?

Quite a lot it tuns out.

Pertti just happens to be an expert on the two species of birds that most interest me, ospreys and owls. He has done the work we did today, surveying Finland's osprey nests, for more than five decades while also studying his country's owls, and he has found the two species inextricably intertwined. He tells me that at one point 10 percent of the predation on osprey nestling and fledglings was from attacks on their nests by Eurasian eagle-owls. This was not Flaco learning to hunt rats. This was a large bird attacking the nest of another large bird. *Intra-guild predation* is the term for one raptor attacking another.

Those attacks fell as the eagle-owl population did, and the osprey population rose. Many other factors, including the decreased human use of environmental toxins such as DDT, led to the rise in the number of ospreys, but having their homes no longer dive-bombed by eagle-owls certainly didn't hurt. Pertti wrote of ospreys in 2006: "Of the 1526 potential nest sites inspected in 2006, 987 were occupied, 795 were active (eggs were laid), and 744 were successful (young were produced). The latest population estimate for Finland is 1200 breeding pairs." My experience today with Pertti suggests that he inspected all of those 1526 nests by himself. Likely he climbed up to most of them.

Ospreys weren't the only birds he took stock of. He also inspected the nests of other raptors and found that while ospreys were on the rise, eagle-owls had gone through a more complicated and bumpier trajectory. There was a long history of eagle-owl persecution in Finland, with the hunting of the birds not just allowed but encouraged since they were perceived as competing with human hunters. The population started to increase in the 1970s and 1980s once the hunting of eagle-owls was banned. This growth was aided by an unexpected boon that Flaco would have appreciated: "the excellent year-round food supply of Norwegian rats." The rat boom occurred partly due to the abundance of loosely run local dumps where the rodents feasted, covering mountains of trash to the point that, as Pertti told me, the rubbish hills looked alive, throbbing with rodents. The eagle-owls thrived on this easily accessed food source, just as Flaco had. But since the mid-1990s many of the dumps had been closed which, according to Pertti, "is certainly one of the factors behind the steep negative trend" of the eagle-owl population.

I consider these connections as I drive my rental car south toward Helsinki. Maybe *consider* is too mild a word. I am dazzled by them. I return to the old John Muir quote about picking up one thing and finding it "hitched to everything else in the universe." Which suggests a corollary: pick up two things and you will inevitably find them connected.

The connections continue to dazzle the next day. Finland had famous urban eagle-owls long before the United States. My hotel is by the harbor, and it is just a short walk down the Esplanade to the Hotel Torni, the spot where eagle-owls once perched high above the city. I ride the elevator to the twelfth floor and then climb a circular staircase to the rooftop bar. Helsinki is a beautiful city but a low one with few tall buildings, and this is one of the highest points in the city, 69.9 meters to be exact, with a view of all of Helsinki. The bartender, whose name is Gere, suggests I try the local lager called Karhu, which means "bear."

"Karhu." I say the word out loud a few times. By memorizing it I am effectively doubling my Finnish vocabulary, which at the moment consists of one word: *huuhkajat*. Like a child, I am starting with animals. Karhu. Huuhkajat. Bear. Eagle-owl.

Gere himself looks slightly bearlike, burly and redheaded, wearing traditional colored suspenders, and he tells me a little about the history of the hotel, how the Russians took over the building right after World War II.

"They wanted it because they wanted to look out over all of Helsinki, all of Finland," he says.

I tell him eagle-owls perched here for exactly the same reason.

I am surprised that he doesn't know about the history of the urban owls, which Juha has regaled me with, but he only started working here two years ago, and for the most part the owls have disappeared from the city.

Soon I am looking out through the floor-length glass windows at the same vertiginous view that the Russians and owls once had, the city spread below me, the harbor and the Gulf of Finland and Baltic Sea to the south. You can see immediately why the eagle-owls favored this place. Like the sheer cliffs that are their native habitat, the building allows you to take in everything below, with the city laid out like a map.

This city was infatuated with urban eagle-owls long before New York was, and from up here I can see many crucial spots in the Helsinki eagle-owl story. Just to the north are the blue glass squares of a building that features its name—FORUM—in huge letters near the top. From photos I have seen, I know that an eagle-owl used to perch atop the *F*. Over to the east is the Kallio Church, 308 feet high and sitting atop a hill with a gray granite facade and a cross at the top that is iron oxide green. The outgoing receptionist at my hotel told me that her boyfriend has seen owls atop that church, and tomorrow I will walk over and inspect. The tower will be closed, but I will sit alone in the pews of the cavernous church, which can hold 1,100 worshippers, and think that, with its lit candles and the beautiful mosaic, it is the right place to dwell on the sacred aspects of owling. Three hundred feet above where I sit the squat cross atop the church once served as an owl's pulpit.

It is a mostly clear day, with a few cloud continents shifting above, and to the south I can see the islands that dot the harbor. Further south, out of view today, is Estonia. Swifts and a few gulls fly by. I order a second Karhu, chat with Gere, talk to the people at the neighboring table about the owls.

You never know where your obsessions will take you. Mine have taken me to Finland. I think, *How did I get here?* Why *did I get here?* The answer to the second question is simple: huuhkajat.

Further to the north, past the Forum, is a dark line with flags flying above it, that Gere tells me is the Olympic Stadium. The stadium was built for the 1952 Olympics, with Helsinki being the northernmost location for the summer games. The stadium, rebuilt in 2000, might be the true church of the owl.

"The world's most famous bird," is what people started to call Flaco almost from the beginning. But who you call "most famous" depends on where you live. In Finland, like New York, many claim that title belongs to an eagle-owl, but here they are talking about Bubi, not Flaco. Bubi is the name Finns gave to the eagle-owl, a play on its Latin name *bubo bubo*. Bubi became famous by flying into the Olympic Stadium during a qualifying match for the European Football Championship in 2007. The Finnish national team was playing Belgium, who they rarely beat. The giant bird swooped down and alighted atop the Belgian goal, forcing the referee to stop the game. Fans cheered and heckled (since it was Finland these were gentle and relatively polite heckles). The owl took its time, no doubt staring down the players with its orange eyes, then flew to the goal at the other end before flying off to more roaring cheers. Finland won 2–0.

After that the Finnish men's team was called Huuhkajat, a name they have kept ever since. The women's team would eventually have their own nickname, Helmarit, which means "boreal owls."

It was in the sauna in the basement of Juha's house that I first heard the full story of Bubi. Juha's brother-in-law, a well-known Finnish composer, told it to me as sweat dripped from my brow.

"Now *that* is the world's most famous bird," he concluded.

JOHN MUIR WAS RIGHT.

Find something. Pick it up. Follow it. Care about it; maybe even obsess over it.

If you do, you might find that your life widens and maybe even deepens. And you might find that it leads you to unexpected places.

You also might discover the difference between what the philosopher Alfred North Whitehead called "inert ideas" when contrasted with *living* ideas. The first are facts and ideas that are understood in a merely intellectual or academic fashion. The second are living things, tied to our concerns, enlivened with the electric jolt of an uncovered wire.

Back in the spring I taught an environmental writing class for my graduate students. These were mostly students who specialize in nonfiction, and they tend to mostly write memoir. While I admire their devotion to the first person, a devotion I share, I always try to nudge them toward writing about the world. At the beginning of the term I gave them an assignment, one that my own teacher, the Native American poet, essayist, and novelist Linda Hogan had given me thirty years before. The assignment was simple: Pick an animal, any animal. Go and watch it. Then study it, learn about it, talk to experts about it. Immerse yourself in its world. Students picked alligators, jellyfish, dolphins, manatees. Some good writing came out of it, I think. At spring break I suggested they go on journeys following their animals. I told them that these might actually be more enjoyable than sipping fruit drinks on the beach. I called these trips "vacations with a purpose."

It seemed to me that many Flaco followers, despite not having been assigned it in a class, were already doing this sort of homework. The unusual thing, however, was that Flaco was a group project. The traditional nature writing quest, from Thoreau at Walden to Muir himself, is a solitary one, taken most often by questers of my gender and skin color. While I still value solitude and feel that it offers insights and ways of thinking apart from the crowd, I am intrigued by the possibilities of a community quest.

The community has not dispersed, even six months after Flaco's death. It's true that some Flaco followers have sworn off birds, too saddened and hurt by the death of *their* bird. But for many others Flaco continues to provide a kind of portal into a different world. Nan, whose life has clearly been altered by Flaco's visitation, has spent the better part of her summer working on the children's book about the owl. David Lei and Jacqueline Emery have, like me, flown across the Atlantic to learn more about eagle-owls, first studying the urban birds in Spain and then, later in the summer, flying to Germany to see them in the wild. Flaco meanwhile remains very much alive on X, where David Barrett keeps the flame alive, and on Facebook, where a woman named Kathy Robles posts daily photos from his wild year, many of them never before seen by Flaco followers. The angry shouts about bringing the vandals to justice have died down, and the community carries on. People still grieve, a word that might sound too strong but that I'm convinced is not. Not long before my trip overseas, I attended a get-together in a studio in New York, hosted by Dr. Khan from our Zoom group, who read a poem about Flaco. Other attendees included Anke and Bangkok Dave, and everyone shared their photos and artwork and writing.

FINLAND IS TURNING ME into a night owl.

Like David Barrett before me, my hours are shifting due to the animal I am pursuing, but it isn't only that. At home I am an early bird, but here we are seven hours ahead.

After my visit to the top of the Torni Hotel, I take a long nap, and when I wake, groggy and uncertain where I am, it is almost seven. At home my day would be ending. Here I take the five-minute walk to the harbor and decide, impulsively, to get on a ferry. It is a beautiful night, the Baltic Sea calm and the sky clear, and soon I am heading to the island of Suomenlinna.

Before I leave Finland there is a mystery to be solved. Why

are eagle-owls scarce in the city? Recent literature tells me that not long ago there were as many as seven nesting pairs. The population decline of the species in the countryside makes sense. As Pertti said, the closing of the local dumps meant no more rat buffets, and so the numbers have fallen. Development also has had its impact, though from what I have seen much of the Finnish countryside remains good eagle-owl habitat.

Eagle-owls are territorial, and the young, once they finally leave the area where they are born, need to claim territory of their own. So perhaps the boom in the country pushed some young owls into the city, and this ended with the rural boom. Perhaps. I have reached out to everyone I can think of, including birders and scientists and experts at the Helsinki zoo. I've learned little, though somewhere in my notes I wrote down the word *herpesvirus*. Perhaps the reason the heyday of the Helsinki owls ended is the same reason that Flaco died. Perhaps their story is his story, and urban life—the viruses, the rat poison, the collisions with windows and vehicles—has proved too much. Somewhere in Pertti's papers I also read about "owl electrocutions," and it turns out that many owls die when landing on or colliding with power lines. We have made life hard for them.

The ferry is pulling up to a green island spotted with buildings and homes, radiant in the autumn light. I have seen my eagle-owl in the wild, but now I want to see a bird like Flaco, one who has adapted to life near humans. The other day Juha told me that there might still be eagle-owls nesting on one of the islands just beyond the harbor. At first he believed there might be a pair on Vallisaari Island, home to a military base until recently and known as "rampart island" for its old forts. I considered kayaking out to it, though I was warned that the shipping lanes were dangerous. But this turned out to have been an outdated report; there were no longer owls on Vallisaari. There was a rumor, however, that there might be an owl on this, the larger island of Suomenlinna, though no one knew exactly where. I am on the ferry following that rumor.

When we dock on the island, I have no idea where to go. But I see that there is a small general store not far off, and I follow some fellow passengers in that direction. Seventy percent of my job on these trips, and of my working method as a writer, is talking to strangers. I see a guy coming out of the store carrying a long tubelike black case that I think may contain a fishing rod. I approach him on the principle that fishermen know where the birds are. He is maybe in his thirties, with a head of thick dark hair, and it turns out the case actually holds ski poles for hiking. His English is pretty good, and he says that though he hasn't caught sight of one this year, he has seen an eagle-owl here in years past. He suggests they might be nesting over by the old shipyard—as he searches for the exact word in English someone else coming out of the store stops and provides it for him: dry dock. He is trying to explain where the dry dock is, and I know there is no way I'll be able to find it, when another man, this one older, says he is going that way and can take me there.

Eric wears a stubbly white beard and a bucket hat and has a little dog of indeterminate breed who he picks up and carries as we walk. We amble over cobblestones past an old church while he tells me he spent his life at sea before retiring to Suomenlinna. He says he reads the *Guardian* and watches CNN, and I get the sense, not for the first time on this trip, that this country might have more at stake in the coming U.S. election than we do due to their history of being invaded by Russia and their fear that they will be next on Russia's plate after Ukraine. Eric says we need to stop at what he calls his "cabin" to drop off the dog and then insists on giving me a tour of the place, including the sauna.

Back on the road we walk over a beautiful low-arching bridge and turn right toward what I now see is a boatyard. We don't pause there, however. There is a small restaurant past the boatyard, and Eric thinks someone in there might have an idea about where the eagle-owls are. When we look inside, however, we see that most of the waiters and the maître d' are very young, teenagers, or maybe in their early twenties. We share a quick comment about the unlikelihood of these kids being able to help us with a bird question, assuming that their preoccupation with their phones and other teenage obsessions would preclude it. This turns out to be generational prejudice. Elias, blond and shy and likely the youngest of the waiters, says he knows a little and leads us back outside toward the shipyard.

"My coworkers told me about them," he says. "I have seen only one. Earlier in the summer. It was up there."

He points up at the construction crane that looms above us.

We thank Elias and return to the dry dock. Fencing blocks off the area, clearly marking it as off-limits, but Eric suggests I ignore it. He says he will do the same; he is interested in picking up some stray boards for repairs on his cabin.

The area we walk through is cluttered with scaffolding and two-by-fours, parts of ships, ladders, stanchions, cement blocks. Twenty yards in there is a twenty-five-foot drop-off into a kind of graveyard of old ships in various states of disrepair, none of them looking like they would float. I stand on the edge of this area and look across the chasm at a massive Swedish fort, built in the 1700s and later captured by the Russians. A hundred yards long and forty feet high, its walls are a mosaic of stone topped by a row of brick. What catches my eye immediately are the series of alcoves that look like rough bay windows, but will turn out to be gun ports, along the top. There are ten of them, black caves maybe six feet across, thirty feet up. I glass them, thinking that if I were an eagle-owl that's where I would be. I see what might be a log or stump in the third alcove from the right.

While I stare, Eric scavenges for lumber. When he returns, he starts telling me about his life on the sea. It is interesting stuff, but now I have eagle-owls on my mind, and I am not unhappy when he at last decides to leave. I stare at the log in the third bay. Down below me, in the chasm of lost boats, are eleven barnacle geese, with black heads, white chests, and gray wings, who wander over the cobblestones between the ships. When I return my gaze to the third bay, the log is looking more owlish, but that could easily be my hungry imagination. And then the log moves.

I am excited but not entirely certain what I am seeing, but gradually the log/owl starts to clarify itself. At nine o'clock the bells in the church start to play. The light is slowly fading. Through my binoculars I think I can make out the ear tufts and the slashes of black on the buff chest. But I'm still not sure. I need to watch this not-yet-owl carefully in case it moves again or possibly flies. I don't allow myself to even look down at my phone. Time passes, maybe an hour. And then suddenly the owl fully announces itself, walking down out of the depth of his alcove to its outer edge, the doorstep of the bay. I can hear his footsteps on what I assume must be an aluminum floor, and I now notice the debris—old food and bones, the innards of pellets—that litter the edge. The owl stands right on the edge of the bay, his talons gripping it, looking ready to push off, but he just scratches himself, turns around and walks back into the deeper cave of the gun bay.

Another hour passes. Real darkness starts to creep in. At one point a man in a sleeveless shirt with tattoos on his arms starts walking through the fenced-off area where I have camped out. Someone I might try to avoid in other circumstances. But when he gets close, I point up.

"There's an owl up there," I say, not taking my eyes off it.

"It's always there," he says in decent English.

It turns out he works here.

"He's our dockyard mascot. We call him Mr. Who."

He leaves and I keep watching. I recognize this as a test of patience. I will not miss flyout. Another thirty minutes passes, and real darkness begins to settle. Mr. Who finally decides to stroll back down to the edge of his alcove. He moves his head, bobbing and shifting, listening now, zeroing in on sounds. This is his launching pad, and he is not going to push off until he is really going after something. Though his hunt will culminate in a dramatic violent moment, eagle-owls do a lot of this. Much of what they do is wait. Wait and listen and watch.

Waiting and watching were also at the core of the Flaco phenomenon. We can look at the whole thing through a glass darkly and see it as a convoluted postmodern story created by social media, one that sent small armies of people chasing a beleaguered bird. But we can also see that at its essence it was about human beings watching an animal. To follow an animal. To watch an animal. To learn about an animal. It is an experience encoded in humans from our very beginnings. To quietly watch the creatures we share the world with. One of the very first skills of our species, one of the first vital tasks we took on. No wonder if feels so good to do it.

For the most part I resist reaching for my phone, only taking a couple of pictures of the old fort. Photography was central to the Flaco experience. On the one hand there is something distinctly un-present about our mania for picture-taking, the way we instinctively reach for our phones like gunslingers, that is, if we ever put them down. It says something about our society's hunger for being other than where we are, for wanting to remember a thing even before it has happened, for treating moments like trophies. But to pursue a perfect picture, like Anke or David Lei, is different. It is to be a hunter, and to be a hunter requires virtues quite different from those valued in our fast-twitch world. It requires waiting, not moving, not readying for the next thing. Which means that photographing owls, or simply watching them, requires the exact skill set that owls have in spades.

In this light the big bird looks almost tiger striped, his marking distinct and his eyes orange. I try to tamp down my excitement. He seems poised to fly, and that is how I feel too: poised. When he pushes off at last, I want to cheer, but he does not dive. He merely flies up to the roof, perhaps to get a better view. I can hear him now. Not a hoot but a kind of curt *eh-eh-eh*, like a cough almost.

Finally he drops off the fort, his wings spread, and swoops down into the boatyard chasm, taking aim at the barnacle geese. He pulls up before hitting a goose, and though they squawk I am surprised when they don't fly off, instead just waddling away between the boats. The owl lands on a block of wood not forty feet from me. I can see how big he is now, can see the blazing eyes up close. He is very aware of me and stares right back. He is right on top of the geese, but they do not fly off, even when he swoops down again. Like he is playing a game.

A security guard approaches, but I try to keep my eye on the owl.

"You can't be in here," he says. "You can't be behind the fences."

I tell him I am here to watch the owl, that I have come from the United States to see it, and soon he is forgetting about ordering me to leave and telling me about the history of the fort and the history of the owls. He is the one who informs me that the owl's home is an ancient gun bay.

"They have been living here for quite a long time," he says.

His accent is thick. Russian, I think. No, Finnish, he corrects me, offended. We watch together as the owl takes another swipe at the geese and then flies off, landing atop one of the construction cranes. When he flies off the crane, I lose sight of him.

The security guard moves on, and I stay for a while, hoping the owl will come back. When he doesn't, I leave the fenced-off area and walk over toward the crane. It is true night now, and I think I had better leave, but then I catch sight of the owl atop a fir tree. I watch him, a black silhouette against an almost-black sky, until he flies back atop the crane.

It is past midnight now, and the island is empty. The ferries run late, but I am not sure of the way back since on the way here I simply followed Eric. I walk up to the church, uncertain and lost, and discover a couple, their arms around each other, walking across the cobblestones. I head over to them and ask them where the ferry landing is. They have no English; in fact we have no words at all in common. Finally the man uses the translation device on his phone and shows me the screen. One word: *Boat?* I nod and he points me down one of the streets running away from the church.

I am almost alone on the ferry back. We stream across the harbor toward the glittering lights of Helsinki.

THE NEXT DAY I pay a somewhat obligatory visit to the Helsinki zoo. It is a nice zoo with large cages, located on an island all its own à la Jurassic Park, and yet I quickly get that sinking feeling again. Especially when I see the haunches of a wolverine clambering up through its playpen or the delicate feminine yellow eyes of the snowy owls looking out from the lush green of a nearly tropical habitat that offers them neither camouflage nor comfort. The eagle-owl cage is bigger than the others I have seen, more racket-ball-court-size than department store window, and I have been told that the captive bird was injured and would not survive in the wild, but still I am relieved when I can't find it in the cage, either because it is missing or too well hidden.

Before I left on my trip, I paid another visit to the Sylvan Bird Sanctuary where Flaco was born. There I got a glimpse of what Flaco's first year must have been like, before he was moved to New York. What I had not realized when I visited two and a half months before was that the female had laid, or was about to lay, an egg. On my return there were three owls in the cage, the youngster fixing me with an intense stare with its cartoon eyes. Undeveloped ear tufts and fluff balls of down surrounding the facial disc marked it as a young bird. A female, she was already larger than her parents, having gone through a stretch during her first month when she gained between 10 and 30 percent of her body weight every single day.

I learned these facts about the fledgling from Dustin Foote, who told me to call him Dust. Dust was the director of aviculture at Sylvan. We had walked down together to the owl cage.

"We let the parents do the rearing," he said.

This makes sense. Its parents would be feeding it in the wild, and though the birds are territorial it would be kept with its parents until what Dust called "the next cycle," which I took to mean the next spring when the breeding cycle would start again. That would have been when Flaco would have been taken from the cage, coming into his second spring, and shipped to New York.

"Breeding is a very natural behavior, so we try to promote that," Dust said, though he didn't say how, or if there were any owl aphrodisiacs involved.

I found myself wondering why the Central Park Zoo didn't provide Flaco with a mate. Was it simply that the cage was too small? I don't know.

Dust was wary around me, as well he should have been. When I brought up the Central Park Zoo, and he said, "They were definitely in a no-win situation," he might have been speaking about zoos in general. Zoos are a business, and every year that business gets harder and harder with more and more red tape and with the public perception turning darker. Efforts to create larger and more natural habitats need to be balanced by the simple fact that if people don't see the animals, they won't be coming back.

"I have a lot of respect for inner-city zoos," Dust said. "Most inner-city kids would never have a chance to see wildlife without them. The whole point is getting kids face-to-face and connecting with animals so they care about them."

As for the Wildlife Conservation Society he said, "They do so much amazing work that just got shoved to the side. Some of the original pioneering work to turn around hoof-stock populations and bird populations were financed and supported by the WCS."

I remember the sheer delight my daughter took in seeing the elephants at the National Zoo in D.C.

I also wonder if we have reached a tipping point.

Around the time I pushed off for Finland the Oakland Zoo decided to send Osh, its last remaining African elephant, to an elephant sanctuary in Tennessee. The decision was made by the zoo based on the elephant's "well-being," and in a statement the zoo said: "He will have the opportunity to socialize and develop relationships with many other elephants over his lifetime. Something that we could not offer him here."

More specifically it would give the thirty-year-old male an opportunity to reunite with Donna, who had been the last female elephant at the Oakland Zoo before being transported to the sanctuary a year before.

At the sanctuary the elephants move in herds. The sanctuary is a sprawling 3,060-acre habitat, located 85 miles south of Nashville. Though elephants are given individual care when needed, they wander freely on the property. The elephants are not managed: "Recognizing that elephants are wild animals with complex physical and social needs, there is no free contact management or dominance training." The sanctuary literature does not call the elephants "tame." It calls them "captive."

The herd can be viewed on camera, but the elephants' habitats are closed to the public. The sanctuary says this explicitly on its website: "Visitors to The Discovery Center will not see or interact with elephants."

The sanctuary is for the elephants, not the watchers.

ON MY LAST NIGHT in Finland I return to Eagle-Owl Hill.

Juha will join me later, but I asked for a couple hours alone. The problem is that, alone, I can't find the spot where we cut in to the path up the cliff. I wander below the granite walls through the woods for an hour, looking for a way up.

The walls echo with the cries of hungry juvenile eagle-owls. As a connoisseur of wild experiences, there is a certain disoriented flavor to this one: being lost.

I finally make it back to the granite ledge and am almost immediately greeted by the owl. The second visit does not have quite the same magic as the first, unable to match the mix of surprise, expectation fulfilled, and sheer novelty, but over the next two hours I will witness a variety of interesting behaviors as the owl is dive-bombed by a sparrow hawk, takes several flights right below my granite stage, and flies back and forth between the twin towers of spruce and the firs closer to me. During the years they were hunted in this country, eagle-owls learned to be wary and avoid humans, but when we changed our behavior, they changed theirs, and this owl, perhaps habituated by years of human visitation, is not afraid of getting close to me. When it does, I see and feel its sheer size and power.

I am encouraged by the fact that both humans and owls can change our behaviors. When we managed to stop poisoning and hunting them, eagle-owls grew less afraid of us. It could be that this owl, easily visited, has been sometimes fed by humans over the years, which might be why it now comes so close. Or it could simply be checking me out and waiting for me to leave. I don't know.

In *The Abstract Wild*, Jack Turner focuses on one particular definition of wildness, a definition that Thoreau made note of; the fact that *wild* is "the past participle of 'to will': self-willed." Turner argues that wilderness can be defined as "self-willed land" and that wild creatures are similarly self-willed. What does this mean exactly? It means that wild creatures tell their own story. The story that this owl has fashioned up on this craggy rock ledge—mating every spring with the same female, fathering multiple broods of young, thirty-nine juveniles flying away from the hill to start their own lives, getting to know every nook of its territory, learning the patterns of its prey down below, enduring the heat of summer and the cold of winter—is all its own.

We can argue all we want about the meaning of Flaco. Yes, he was still a captive bird, tame in some ways. But by Thoreau's definition his final year was a wild one. The story he wrote was his own. During that year he changed and grew and learned. He acquired new skills, or rather old skills newly found. His was a tragic story, but it was his. For the first twelve years of his life everything, from what he was fed to how far he could fly to whether he could interact with other birds to finding a mate, was controlled by human beings. In his final year that was not the case. It is true that his year was influenced by man—man's poisons, man's buildings, man's attempts to capture and watch him—but he was wild in this sense: he authored his own story.

I turn away from the bird and thoughts of the bird and focus on my journal and, inevitably, on myself. I write: *What, if any, is the deeper meaning of encounters like these? Can I honestly use the word "sacred" and if I do what does it mean? At the very least these moments seem capable of jarring us out of our own lives, our own enclosures.*

A shadow passes over my journal page and then over the granite and lichen-covered rock below my feet. I look up to see the owl flying right above my head.

A few seconds later Juha arrives, and I laughingly try to explain what I have just experienced. Juha and I have grown close during the last week. Over the next hour we watch together as darkness falls. The owl, back in the taller of the two spruce trees that Heikki called "the twin towers," seems to watch with us.

The light fades, the young still hectoring their father. He huffs up his wings, changes posture. It is time once again to go to work. He seems to stamp his feet on the branch he is perched on. Finally he pushes off, spreads his wings, and we watch as he swoops down into the valley below.